CORE

CORE

A Science-Backed Approach to Exercising and Understanding Our Central Anatomy

OWEN LEWIS

Chichester, England

North Atlantic Books

Huichin, unceded Ohlone land
Berkeley, California

First published in 2024 by
Lotus Publishing
Apple Tree Cottage, Inlands Road, Nutbourne, Chichester, PO18 8RJ, and
North Atlantic Books
Huichin, unceded Ohlone land
Berkeley, California

Illustrations Amanda Williams
Text Design Medlar Publishing Solutions Pvt Ltd., India
Photographs Courtesy of Alamy or Shutterstock unless otherwise indicated
Exercise Photographs Joel Hicks
Exercise Models Madelaine Winzer and Scott Broadway
Cover Design Jasmine Hromjak
Printed and Bound in Turkey Kanat Sultur Printing House

Core: A Science-Backed Approach to Exercising and Understanding Our Central Anatomy is sponsored and published by North Atlantic Books, an educational non-profit based on the unceded Ohlone land Huichin (Berkeley, CA), that collaborates with partners to develop cross-cultural perspectives, nurture holistic views of art, science, the humanities, and healing, and seed personal and global transformation by publishing work on the relationship of body, spirit, and nature.

North Atlantic Books' publications are distributed to the US trade and internationally by Penguin Random House Publishers Services. For further information, visit our website at www.northatlanticbooks.com.

Medical Disclaimer: The following information is intended for general information purposes only. Individuals should always see their health care provider before administering any suggestions made in this book. Any application of the material set forth in the following pages is at the reader's discretion and is their sole responsibility.

British Library Cataloging-in-Publication Data
A CIP record for this book is available from the British Library
ISBN 978 1 913088 40 8 (Lotus Publishing)
ISBN 979 8 88984 071 8 (North Atlantic Books)

Library of Congress Cataloging-in-Publication Data
Names: Lewis, Owen (Personal trainer), author.
Title: Core: a science-backed approach to exercising and understanding our central anatomy / Owen Lewis.
Description: Berkeley, CA : North Atlantic Books, 2024. | Includes bibliographical references and index.
Identifiers: LCCN 2023033930 (print) | LCCN 2023033931 (ebook) | ISBN 9798889840718 (trade paperback) | ISBN 9798889840725 (ebook)
Subjects: LCSH: Physical fitness. | Exercise. | Muscle strength.
Classification: LCC GV481.L615 2024 (print) | LCC GV481 (ebook) | DDC 613.7/1--dc23/eng/20231106
LC record available at https://lccn.loc.gov/2023033930
LC ebook record available at https://lccn.loc.gov/2023033931

Contents

Foreword

In your hands is a book that only Owen Lewis could have written.

When I first met Owen, he was a keen student in a bodywork training I taught in England. He stood out from the rest of the class, not just because of his height but because of the questions he asked. Owen was always searching for definition and meaning, for clarification and interpretation; he was never content with a glib or superficial answer.

During breaks and after class, I often found Owen with his head in a heavy tome during any quiet moment. I discovered Owen devours words. He is the only person I know who has read *War and Peace*, not once, but numerous times. Owen loves philosophy, music, art, and all genres of literature. The short, digestible book you are about to read demonstrates that those hours of study were well spent as he has weaved those passions together in this powerful and essential text.

After completing his studies with me, Owen sought to quench his thirst for understanding by spending time with other top-level manual therapy and pain science teachers. He came back heavily loaded with but by no means overwhelmed by information, insights, and interpretations from the world's best thinkers. But Owen does not limit himself to thoughts on manual therapy and anatomy. Owen is an explorer of the mind as well as the body, always looking for deeper meanings and better ways of thinking, and finding methods through which we can gain definitions of complex concepts.

Previous authors have presented their definitions of the "core," but Owen is not happy to retread their paths. Instead, he is content in the "not knowing." To define the indefinable core, Owen references Jules Verne, the laws of Hammurabi, Plato, and Thomas Merton, among so many others. Great authors, artists, and philosophers provide us with tools to see the world in new and valuable ways. They provide words and images that bring metaphors and similes to our consciousness and offer us fresh visions—in this case, of what the "core" might be.

All the greatest books give us new perspectives; they enlighten us, challenge us, and change how we see. In its few pages, *Core* achieves all of those things.

I recommend this book to anyone wishing to understand how the body moves.

James Earls
Author of *Born to Walk* and
Understanding the Human Foot

London
September 2023

Introduction

Looking Within

An apple icon glows on my laptop, another bite and it might become the elusive still point in my turning world: the core. The core is a common term we all know, use, and overuse, often without appreciating its myriad of interconnecting elements. The apple core is the apple, as the stalk is the tree, as the roots are the soil, as the leaves are the wind. Separating the core from the body, from the mind, or from our social context is an abstraction. The notion that all these fragments exist separately is a delusion. By maintaining this persistent false belief—despite strong contradictory evidence—we delude ourselves and dilute our training. The division of parts is one created by humans for simplicity, which often seems to create conflict and confusion.

Underlying our analysis of the core is the concept of an unbroken wholeness. Context, mind, and matter can be seen not as separate substances but rather as different aspects of one whole and unbroken movement. The only certainty is uncertainty. A fluidity of confusion and a complex chaos, which, when embraced, becomes a strength to use in our search for, and training of, our elusive core.

Taking a break from work I visited the Tate Art Gallery, London. It surprised me to find that the work of one of my favorite artists, Kenneth Noland, had changed from the last time I visited. His painting, *Turn*, had become the cross section of a torso (figure 0.1).

Clearly, becoming engrossed in this subject had changed my perception. And for Noland, "… context is the key—from that comes the understanding of everything." (Noland, 1988) And while "understanding everything" far exceeds my capabilities, I agree that context is key. Appreciating the core in differing contexts allows one a greater understanding, appreciation, and applicability to movement. It is hoped that this enables the artistic creativity necessary to exercise, as the painter uses the paint brush as a vehicle for physical and self-exploration.

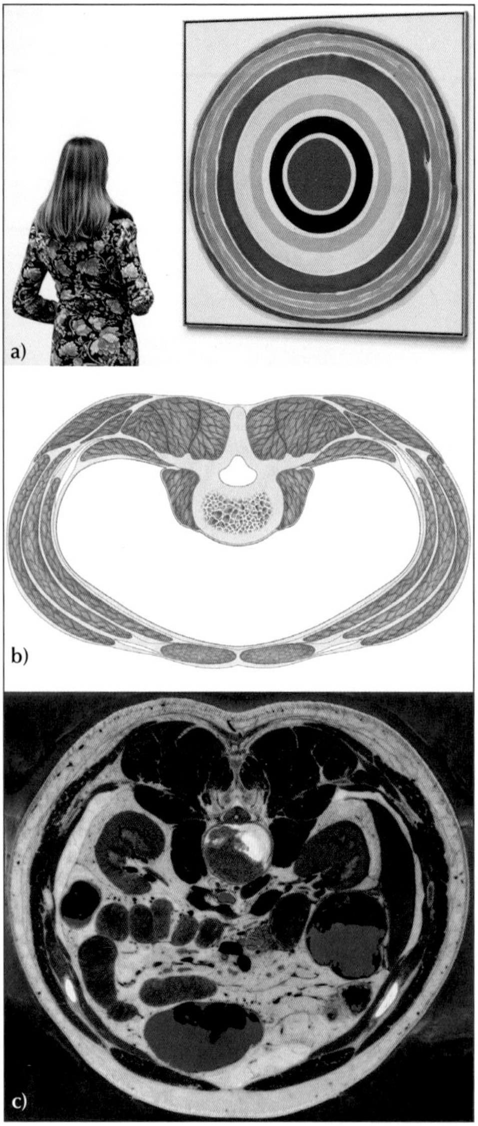

Figure 0.1. (a) Kenneth Noland painting, Turn, 1977; (b) anatomical drawing; and (c) dissection of the cross section of a torso. Each lends a different perspective to our understanding.

As the "Barefoot Professor" Daniel Lieberman explains, many of us are "exercised about exercise" (Lieberman, 2021). Exercise and movement have become medicalized and politicized into various *controversial* topics. It seems fitting that at the center of these controversies, fake news, and quackery is the core. Our understanding of the core reflects "the contemporary, industrialized approach to exercise [that] is marred by misconceptions, overstatements, faulty logic, occasional mistruths, and inexcusable finger-pointing." (Lieberman, 2021).

The core, it seems, is everywhere and nowhere. For some, core exercises represent the miracle cure capable of treating a variety of conditions from back pain to depression and incontinence. For others it is a figment of our imagination that cannot be trained or identified. Whatever the truth, the idea of the core is here to stay, and so it seems a good idea to explore this concept, to embrace and dispel some of its myths.

A myth can be defined as something that is widely believed but inaccurate and exaggerated. So often I find information and ideas of the core that fit this definition of a myth. With so many different myths, definitions, perspectives, locations, and explanations, training the core does, at times, seem impossible. It seems the term *core* is as elusive as it is fashionable—as changeable and uncertain as its indefinable nature. However, as I shall suggest, uncertainty is the essence of our core. Understanding our core's unstable characteristics enables the various practicalities of training. This book is not one of solid certainty or a concrete truth, but intended to be an exploration that starts from the center only to return to the center and so to "know the place for the first time," (Eliot, 1941).

For any journey, there is a necessity to begin with clarity. Our guide to gaining a wider perspective, understanding, and appreciation of the core is the biopsychosocial model. The biopsychosocial model was developed by George Engel in 1977 as a model to better understand medical conditions (Engel, 1977). Engel took a global approach to medicine, realizing that medical issues were not simply due to the biological but were also related to psychological and social factors. This approach gives us the foundation for understanding core training by considering its wider psychological and social context. This book aims to extol the merits of an integrated approach by using this biopsychosocial approach as our guide to explore, learn, and exercise our core; to work out the complexities of working out from within.

CHAPTER 1

Theory and Insights

Define the Indefinable

In 1864, Jules Verne took us on a *Journey to the Centre of the Earth* in his book to discover the Earth's core (Verne, 1864). The human core conjures up similar ideas of journeying to our soft middle. However, the core of popular understanding is not the apple core or Earth's core that represents our center. The human core, as it is most often referred to, is actually more like the human equator: the core muscles that circumnavigate our center.

If we stopped here at this superficial, core equator we would, I feel, fail to allow the possibilities for a deeper exploration as to what lies at the core of our being. This "who am I question" is one that has been at the heart of human curiosity for all time. This central (core) question has driven an abundance of research, philosophies, and mythologies from diverse corners of science and literature.

Perhaps it is the variability of such divergent genres that has led to the lack of any agreed single definition or single theory of the core. In spite, or perhaps because of these many articles, research continues to be carried out, and books continue to be written on the subject. Practically, physical trainers and gyms persist in using, and at times, abusing the term core. Core has become a vague and consistently inconsistent term. The result of this confusion can be seen in the varied forms of training said to work out our undefined core.

Having no clear definition of the core could be seen as a concern, even a weakness, in the theory and its practical application. Struggling to define key terms is not unusual in the exploration of concepts central to human understanding. Similarly, difficult-to-define concepts include the terms: *knowledge*, *information*, *order*, or even *life* itself. I suggest that the inconsistency in defining the core tells of an important underlying story. This story is at odds with conventional ideas of a solid stable core. Central to this narrative are concepts of *inconsistency*, *uncertainty*, *adaptability*, and *fluidity*. These terms seem to reflect the real nature of our physical core. Perhaps it is the transformative nature of our core that has accommodated the multitude and varieties of exercises, as well as changing theories, definitions, and ideas of our core.

Humans have long sought to make life meaningful by securing their understanding into rigid forms, laws, and fixed ideas. The earliest written laws of the Babylonian societies (1792 BCE) were carved on huge stone slabs proclaiming the Hammurabi laws were to be obeyed. Throughout history, societies have readily accepted, and perhaps even craved, for laws to guide us on our transient path through the cosmos. Often it is this elusive search for certainty and clarity that led others to realize a different truth expressed in the abstract. Music, poetry, dance, and sports can give the "soul to the universe, wings to the mind, flight to the imagination and charm and gaiety to life and to everything" (Lubbock, 1887).

It is to this more abstract notion of our bodies that we shall use as our inspiration to explore our core. Our bodies are not fixed stone structures bound by law or convention but are moving forms of abstract expression. Rather than attempting the impossible task of making fixity out of flux, I attempt to embrace the mobile, adaptable abstract form that is the human body and, specifically—the center of our abstraction—our core.

As a result of my research for this book, my conclusion is that the core is both nowhere and everywhere. "At the centre of our being is a point of nothingness which is untouched by sin and by illusion, a point of pure truth" (Thomas Merton, 1966). This realization is one that may create the potential for confusion rather than a "point of pure truth." This is especially true as we begin to discuss the anatomy of the core. It has been argued that the core musculature should include the muscles in the shoulder as it is critical for the transfer of energy from the larger torso to the smaller extremities (Hibbs et al., 2008). McKeon uses the term *foot core* to explain the function of the intrinsic muscles of the foot (McKeon et al., 2015).

From this, the common idea of a central core seems to be an inaccuracy. Many of the concepts that surround ideas of the *core shoulder*, *core foot*, or *core torso* are universal to the body and to our understanding of the body in movement. For the purpose of this book, we shall confine ourselves to exploring the area from the pelvis to around the fifth rib. It seems for expert and layperson, this is the area is most often nicknamed the core. As we shall discover, the character of the core defies definition. I shall not attempt—where others have failed—to define the core, instead the attempt is to use the strength of this undefined concept to explore further.

An Exercise in Philosophy

It is worth recognizing that any training, core or otherwise, should be part of an overall balanced "diet" of training. Just doing the exercises advocated here will not be as helpful as combining it with Pilates, yoga, weight training, running, cycling, walking the dog, DIY, or anything else that moves you to move.

Given that the core is at the center of the kinetic chain and described as the body's powerhouse and the foundation or engine of all limb movements (Akuthota & Nadler, 2004), it is conceivable that (almost) any and every exercise could be labeled a core exercise. This book could have become nothing more than a huge list of exercises of each of the twenty-nine muscles of the core. To allow this book to be of a manageable size, and save a few trees in the process, our aim is to use exercise to illustrate key principles that underpin training. It is hoped that this principles-based approach will allow further development and tailoring of exercises to individual needs. The exercises within this book aim to manipulate the environment to direct forces to alter the demands on the body and, specifically, the torso.

For example, just changing one's position, from standing to sitting or lying, will alter the direction of forces to affect different aspects and requirements of the musculature. Different equipment, such as resistance bands, suspension trainers, dumbbells, and barbells allow a different loading emphasis. At each step, we look for variety by changing the driver of the action, the tempo, and other aspects of the environment to modify training. The exercises shown can, and should, be tailored to meet the specific needs of the individual and the task. The exercises and discussion surrounding them will, I hope, show the importance of variety in training. I intend the exercises to demonstrate the principles to allow a diverse and wholesome approach to training.

Our increasingly sedentary lifestyle has emphasized the importance of movement as the fundamental aspect that defines life. "Life is defined as any system capable of performing functions such as eating, metabolizing, excreting, breathing, moving, growing, reproducing, and responding to external stimuli" (*Encyclopedia Britannica Online*, 2022). In some utopian existence we would all move enough and in the right ways to sustain a fulfilling life. "If we could give every individual the right amount of nourishment and exercise, not too little and not too much, we would have found the safest way to health" (Hippocrates fourth century). However, for the normal mortal who uses cars, computers, and sofas, a little help is needed to move more often and a little better. In our attempt to counterbalance the otherwise inactive conventional lifestyle, society has turned to exercise as its savior.

What motivates us to exercise is a curious and involved question. The answer is individual and multifaceted. For many, the motivation to exercise stems from ideas of "should." "Should" means we exercise from an external drive. "Should" exercisers often move because others say it's good for us, reduces body fat, improves mood, drives away depression, or builds a body that conforms to one of society's fickle fashions.

However, motivated by "should" often means that, given half a chance, we'll skip the workout or miss the gym session. Gyms across the world rely on this to make their money. It is an unusual business model; gyms thrive partially because their customers don't attend. Imagine using this same business model for a shop—fewer customers normally mean lower profits. However, if everyone with a gym membership attended their gym at the same time, the line for the treadmill would be longer than the run one intended. Years ago, wildly overcrowded gyms would have been forced into becoming huge establishments catering for the whole population. Reality indicates that the morality of "should" is a fickle way to motivate us to benefit from exercise.

Progressing from a "should" exerciser requires a little thought and the careful use of goals. Goal setting can change the "should" exercisers into becoming "driven" exercisers. Goals and plans that link well with a love or desire drives one into "having to" exercise. Motivated by a favorite sport, hobby, or holiday by the beach, we endure the gym or long wet winter runs. For me, it was a love of basketball that drove me into my local gym to endure their terrible choice of music.

With time I, like many, started to want to exercise. "Want" is often driven by a rush of endorphins, improved mood, cognitive functioning, and self-esteem all associated with exercise. As a teenager, I enjoyed seeing the physical changes as I gained a more positive body image. I began to welcome the leaden sense of fatigue that induced a deeper sleep. And my cake was not iced with guilt as it was for some of my friends. The benefits of exercising are numerous and well documented.

For many exercisers, this is where the progress of motivation stops. Remaining driven by our goals and the enjoyment of the side benefits of exercising creates the reason to continue. While this is a long way from the early days of "should," the motivation to exercise is still one incentivized by the reward of goals.

There is another stage, a place where exercising is a pathway rather than a destination or goal to achieve. Today, I attempt to use exercise as the medium to know myself a little better. The aim here is not to necessarily become an expert exerciser or teacher, but it is the path of a student. Exercise is an exercise in understanding physical and mental development. For me, and others like me, exercise has become an ever-changing discipline of reflection and response—a path of self-development.

In this way, exercise, in its numerous forms, becomes a window to an inner emotional self, opening deep centers of creativity and thought. This stage in our process of exercising goes some way toward being able to "remove the idea of a goal from the process and then affirm the process in spite of this? This would be the case if something were attained at every moment within this process" (Nietzsche, 1901). Such goal-less exercise, without an endpoint, retains the idea of learning and progressing with a childlike, modest curiosity that continuously fulfills and motivates the developing forms of exercise. Exercise becomes an exploration rather than a destination. In this way, failure becomes a welcomed part of the learning process to be embraced as an opportunity within this philosophy of exercise. This utopian-sounding idea is grounded in the history of philosophy.

The flamboyantly-mustachioed philosopher Nietzsche was a remarkable exerciser and thinker. His chosen exercise genre was walking. Nietzsche walked "for up to eight hours a day" during his most productive years of writing (Gros, 2008). Walking was not a distraction as it was for Kant, another philosopher famed for his walking, and nicknamed, "the Konigsberg clock" for the regularity of his daily strolls. For Nietzsche, walking was "the invariable accompaniment to his writing" (Gros, 2008).

It seems clear that Nietzsche understood the power of physical activity. He certainly seems to predate the idea that "sitting is the new smoking" (an idea credited to Dr James Levine from the Mayo Clinic; Levine, 2014) when he tells us all to "Sit as little as possible" and not to "believe any idea that was not born in the open air and of free movement—in which the muscles do not also revel" (Nietzsche, 1908). Here, Nietzsche reflects many early philosophers who realized the link between movement and a greater understanding of one's self.

The Greek philosopher Aristotle was so famous for his habit of lecturing while walking that his school was named the Peripatetic school. Peripatetic from the ancient Greek word περιπατητικός (peripatētikós), which means "of walking" or "given to walking about." As if to highlight my point, Aristotle's teacher Plato set over half his dialogues in the gym. Philosophizing and training the mind was, in Greek gyms, as normal as it was to work the body.

The nature of Greek gyms reflects some of their ideals just as today's gyms reflect something of our own. Today gyms often portray a Cartesian dualism that divides mind from body. Filled with loud music and TV screens, gyms seem intent on distracting us from mind and so, it seems, from body. Later we shall expand upon some of the advantages of increasing awareness of body and mind. What is clear is that awareness during exercise can enhance both physical and mental health (Mikkelsen et al., 2017). While exercising to music has its benefits, such as to reduce perceived exertion (Terry et al., 2020), occasionally removing our headphones to meet other philosophical gym goers might let us enter Plato's gym.

A Fluid Definition

> *The beginning of wisdom is the definition of terms. (Socrates)*

Perhaps it is a shock that, in a book on the core, we have turned so quickly to philosophy. However, if by core we mean something that is the very center of what it means to be human then, it seems to me that philosophy is an obvious choice of subject. But fear not, once our philosophical foundation is laid, we shall build our monument to understanding and application.

Think of philosophy and we often think of a dry, dusty subject of the mind. This is partly the fault of the seventeenth century philosopher Descartes. Cartesian duality separates mind from body, a reductionist idea that has thankfully been dismissed as a relic of history. Mind and body, like space and time are now understood to be one (just ask Einstein): as we have spacetime, we have bodymind. The idea of bodymind is perhaps more simply summed up as not a separation of factors but of one unified being we call "human." Continue this idea into exercise and one comprehends that understanding one's body is central to developing a knowledge of one's self.

The central purpose of philosophy, as summed up by the earliest and greatest philosopher, Socrates, is to "know thyself." This demand, not suggestion, was so important that it was carved above the entrance to the renowned temple of Apollo, at Delphi (figure 1.1). With these two words Socrates alludes to a major problem common throughout human history: we usually don't spend time getting to know ourselves very well. Socrates advocates the regular, careful examination of ourselves. Many Socratic concepts pivot around the need to investigate, interpret, reflect, and respond. Taking these concepts into exercise may allow a different emphasis that is closer to Plato's gyms of physical and mental health and discovery.

Some of the early attempts to better know one's self led us to look within, literally. The earliest reference to the brain, found in the Edwin Smith Surgical Papyrus, was written in around 1500 BCE.

Figure 1.1. Temple of Apollo (Delphi), Mount Parnassos in Central Greece.

I imagine these early anatomists were rather underwhelmed as they opened the cranium. There was no tiny homunculus controller or spiritual ethereal light that showed our godlike qualities, only a soggy lump of grayish convoluted organic matter. From these dissections we were no closer to understanding self, consciousness, or the workings of the brain.

Taking this same approach to our core reveals other soggy lumps of anatomized tissue. Our anatomical center, our apple core, is composed of a series of slimy, fluid bags—our organs. These organs are kept healthy not by maintaining their stability (a word synonymous with the core) and lack of movement but by the organs' "capacity to be deformed" (Barral & Mercier, 1988). The natural motion of our body and breath helps to squash and squeeze our inner watery balloons. This process massages and hydrates rather like squeezing and releasing the sponge in your bath to refresh with new water. Occasionally, viscera can become stuck and create restrictions. These visceral tethers may force the body to compensate leading to functional and potentially structural dysfunction. Such examples emphasize that "Nature abhors a vacuum, but fears immobility even more" (Barral & Mercier, 1988).

This leads me to conclude that our outer muscular core needs to reflect the watery substances within. A "solid core" of muscles would surely restrict the mobility necessary for our organ health. For this reason, I advocate adaptability rather than rigidity of our muscular core.

After all, it is estimated that we are about 60% water. Water, an almost formless, groundless ground* that the fourth-century BCE Chinese philosopher Chuang Tzu named "chaos." The ability to adjust and respond to the chaotic world within and without reflects the nature of our watery core. Ideally, our structure is "soft and flexible like water" giving our structure a strength that "nothing can resist" (Tzu, 400 BCE). Perhaps this is why humans are at the top of the food chain having evolved to "adapt or perish, now as ever, is nature's inexorable imperative" (Wells, 1945). Our core must bend like a reed in the wind, not snap like the rigid old oak of certainty. An idea that stems from the ancient teachings of the *Tao Te Ching*.

> *newborn—we are tender and weak*
> *in death—we are rigid and stiff*
> *living plants are supple and yielding*
> *dead branches are dry and brittle*
> *so the hard and unyielding belong to death*
> *and the soft and pliant belong to life*
> *an inflexible army does not triumph*
> *an unbending tree breaks in the wind*

* Groundless ground refers to the book, of the same title, concerning the work of Wittgenstein and Heidegger and the transient, fluid nature of reality. It also refers to the nature of the water within us, the gel-like ground substance of the extra cellular matrix.

thus the rigid and inflexible will surely fail
while the soft and flowing will prevail. (*Tao Te Ching*, Lao Tzu, 400 BCE)

In 1993, Rachel Whiteread won the prestigious Turner Prize for Art by filling an entire Victorian house with concrete. She then, somewhat controversially, set about removing the house to leave the inner negative space. Rachel would have great difficulty using concrete to accurately cast our inner space. Our watery body can become our teacher in the lessons of fluidity, adaptability, and uncertainty. It is to that end that our training invites not a solid concrete core but one capable of managing and adapting to the variety of the demands of life.

The Known Territory

If you don't know where you are going, any road will get you there. (Lewis Carroll)

For many years, learning about the body was a largely fragmentary, reductive process—analysis, the breaking down or anatomizing of the body into parts, and learning each part before attempting to reassemble it. While of use academically, this may have limited practical applications. Within the world of physical therapies (physiotherapy, osteopathy, massage, etc.), this fragmented view has led to some well-meaning practitioners working exclusively on the one area of pain. The idea was that a painful, dysfunctional knee (for example) meant that the knee was the problem. The consequence of such a part-by-part, separate view, was to treat the symptoms without even looking for a cause.

This view has limited practitioners of various disciplines for years. It is now acknowledged that the cause of the knee pain could be the foot, hip, or any non-adjacent part of the body. Research confirms that "seemingly unrelated impairments in remote anatomical regions of the body may contribute to and be associated with a patient's primary report of symptoms" (Sueki et al., 2013). Thankfully, things are changing as more and more practitioners find ways to deal with the cause of their patients' symptoms. Within the world of exercise, a similar change of perspective may be needed. Exercise also needs to consider the whole system as one entity, rather than mechanical parts.

The old, fragmented perception of the body led to exercises becoming highly specific. Exercises targeted single muscles, which was wonderful for bodybuilders but less useful for functional, everyday activities. The result—huge biceps and weak backs from sitting on strange machines to work one muscle. Now exercisers are increasingly realizing the importance of the maxim "every day is a whole-body day," rather than separating a leg day from an arm day. The realization that movement is a symbiotic relationship of all elements of the whole individual is transforming today's methods and styles of training.

To some extent it was the discoveries in the field of physics that began to make possible this change in our perception of the body. Physicists realized that the removal of one part from a whole changes the essence of the part itself. As the physicist David Bohm points out, "wholeness is what is real, and that fragmentation is the response of this whole to man's action, guided by illusory perception, which is shaped by fragmentary thought" (Bohm, 1980).

This seems especially poignant when we consider the function of any part of the body in a healthy moving individual. Classic understanding of human anatomy places the structure, not the function, as of primary importance. This is clearly represented by Rembrandt's depiction of the Anatomy Lesson of Dr Tulp (figure 1.2).

The paradox I see represented in the painting is that seven medical professionals and Dr Tulp ask the one person in the room incapable of movement about movement. The history of our anatomical understanding of movement has been gained by asking the corpse rather than active living individuals. Placing movement at the focal point of our understanding of the body allows us to consider the whole person, the whole movement, and the whole structure.

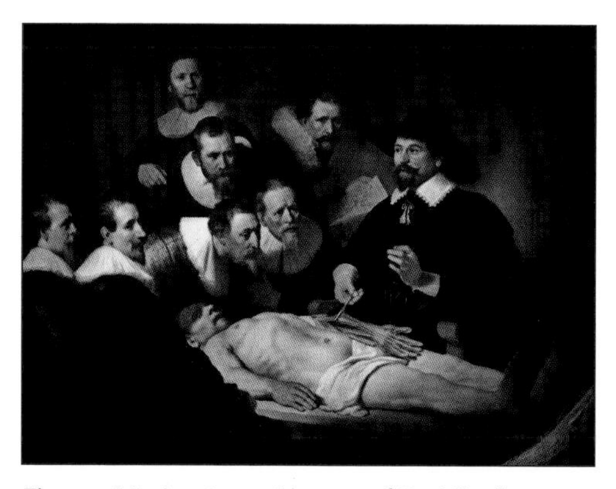

Figure 1.2. **Anatomy Lesson of Dr Nicolaes Tulp** *by Rembrandt, 1632.*

A popular, if general, distinction between Western and Eastern philosophy most often cited is that Western philosophy is fragmentary while Eastern philosophy is holistic (Saranam, 2008). This philosophical idea could be said to be reflected by the differing views of our body.

Traditional Japanese medical and martial arts describe the core as the soft belly or *Hara*. The *Hara* is both a physical area as well as a field of energy. This is like the Chinese concept of the *Dantian*, described as a storehouse of energy in the center of our body. In both descriptions there is an energetic as well as a physical dimension to the core. Later we shall discuss the role of the torso in transferring forces or energy. Perhaps this is an example of Western understanding catching up with some of the ideas of Eastern energy (Beinfield, 1992).

In Western research the core is often described as a box of around twenty-nine muscles, although a definite list is not agreed upon. Some scientists include all the muscles that attach to the vertebrae all the way up to the cervical spine, while others include the deep abdominal muscles like the transversus abdominis (TrA).

While teaching, Thomas Myers developed a concept to explain the interrelationships of our body using a series of interconnected myofascial lines. Within the resulting book, *Anatomy Trains* (Myers, 2020), pictures show how each line was dissected out of a cadaver. With significant skill, imagination, and care, Myers and his colleagues dissected out the Deep Front Line (DFL, figure 1.3).

This is not the usual core of a box around the belly; the DFL functions to lift the arch of the foot (via the tibialis posterior) and runs continuously all the way up "balancing the fragile neck and heavy head atop it all." Conceptually different from the traditional core, Myers still called it the "body's myofascial core" (Myers, 2020). His idea cleverly shows the interconnected nature of our whole system.

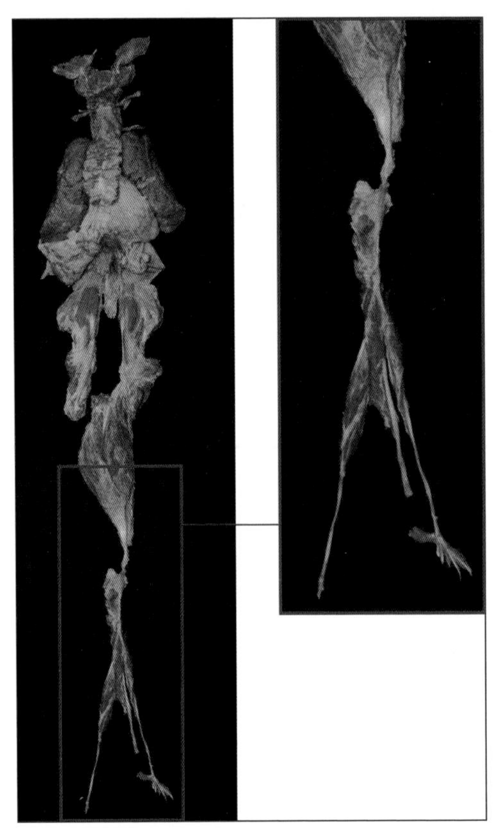

Figure 1.3. The Deep Front Line (Myers, 2020). Image supplied by Anatomy Trains.

Having been privileged to teach within the school of *Anatomy Trains* for a number of years, I found the DFL to be the most difficult line for students to visualize. This difficulty was in part because the DFL should really be called the "Deep Front Volume." A point Myers acknowledges is that the DFL "demands a definition as a three-dimensional space, rather than a line" (Myers, 2020). A degree of poetic license kept the DFL as a line rather than a volume to fit the book's title and inspiration of train lines. While the DFL is difficult to visualize and may not be familiar to the common perception of a core (which looks more like an equator), the DFL is perhaps the closest we have to a fibrous apple core.

> *Beginning at the bottom, the line has roots deep in the underside of the foot, passing up just behind the bones of the lower leg and behind the knee to the inside of the thigh. From here the major track passes in front of the hip joint, pelvis, and lumbar spine, while an alternative track passes up the back of the thigh to the pelvic floor and re-joins the first at the lumbar spine. From the psoas–diaphragm interface, the DFL continues*

> *up through the ribcage along several alternate paths around and through the thoracic viscera, ending on the underside of both the neuro- and viscerocranium. (Myers, 2020)*

Somewhat different from Myers's myofascial core is Joseph Pilates's "powerhouse." Anatomically, Joseph Pilates, of Pilates fame, describes a core that resembles an area most commonly recognized as the core. Pilates's core, or powerhouse, ranges from the pelvic floor inferiorly to the ribcage superiorly. He divided his powerhouse into the five major groups (Muscolino & Cipriani, 2004):

1. Anterior abdominals (also known as spinal flexors). These muscles include the rectus abdominis, external abdominal oblique, internal abdominal oblique, and the TrA.
2. Posterior abdominals (also known as spinal extensors or low back muscles). These muscles include the erector spinae group and the transversospinalis group, as well as the quadratus lumborum.
3. Hip extensors. These muscles include the gluteus maximus and may also include the hamstrings and the posterior head of the adductor magnus.
4. Hip flexors. These muscles include the iliopsoas, rectus femoris, sartorius, tensor fasciae latae, and the more anterior adductors of the thigh at the hip joint.
5. Pelvic floor musculature (also known as perineal muscles). These muscles include the levator ani, coccygeus, superficial and deep transverse perineals, and others.

While anatomists might argue as to what constitutes our core, functionally we acknowledge that no core group of muscles operates independently of all other muscles during activity. "Even the motor neurons of particular muscles are intermingled rather than being distinct anatomical groups in the spinal cord" (Lederman, 2010).

We do not move in a neat muscle-by-muscle linear action. Classification of groups of muscles—like the core—is of use anatomically and theoretically, "but has no functional meaning" (Lederman, 2010). Taking a sip of tea does not require the nervous system to tell the biceps to flex, or the TrA to change its tone, as the pectorals then flexor carpi radialis contract, as there are no biceps, pectoralis, rectus abdominis, or TrA in your brain. The division of us into a series of muscles is a cognitive reductionist construct created by a figment of the dissector's knife. If we really moved in this way, the movements would be robotic, jerky, and the same for all of us. The reality of movement is of complex motor programs specific to both the task and the individual (Hodges et al., 2013).

One only has to watch the Olympic 100 m race to realize the vast differences of performing this same action. Perhaps, most famously, Usain Bolt, at 6 ft 5 in (1.95 m), was deemed too tall by conventional sprinting ideals to run the 100 m particularly fast. After clocking 9.58 sec for the 100 m, it would be tempting to search for the next tall Bolt-like sprinter. However, sprinting greats come in many sizes, at 5 ft 0 in (1.52 m), Shelly-Ann Fraser-Pryce has won three Olympic gold medals and is considered one of the greatest sprinters of all time (figure 1.4).

It is clear that bodies cannot be placed in neat pigeonholes for sprinting or any activity. Another sprinter, Michael Johnson, took 19.32 sec (his 200 m world-record time) to show the world that there is no perfect running technique. Johnson's technique was almost vertical, with staccato strides that, again, went against conventional running wisdom (figure 1.5).

All these sprinters show that that there is "not one strategy of muscle activation that is universally ideal, and not one adopted by all" (Jones & Rivett, 2019). It is hoped that taking a principles-based approach to exercising will enhance each individual's fundamental movement patterns for specific tasks and goals. This book is intended to be a practical guide rather than an exercise in theoretical, anatomical cognition to find any arbitrary correctness.

Despite the inconsistencies, it is widely agreed that training the core (thorax, torso, trunk, or whatever else you would like to call it) is of paramount importance to occupying a functionally efficient

Figure 1.4. Usain Bolt and Shelly-Ann Fraser-Pryce showing that great sprinters come in many different sizes.

Figure 1.5. Michael Johnson demonstrating his unique upright stance while setting a world record in the men's 200 m in 1996 Olympics in Atlanta.

Figure 1.6. A keystone (or capstone)—the wedge-shaped stone at the apex of the arch that locks the other stones and gives strength to the structure.

and adaptable body. At the center of our kinetic chain, the core acts with the pelvis and sacrum: the "keystone of the human body" (Rolf, 1977). A keystone in architecture is the central wedge-shaped stone at the top of an arch (figure 1.6).

The role of the keystone is to hold the overall shape and distribute load. In walking and running, for example, a functional core allows the forces from the ground to cascade up through the torso to the upper extremities. At the same time, bony, myofascial, tendinous, and ligamentous structures transfer and use the force of gravity and twists of the upper body to generate a functional torque to propel us efficiently along on our evening ramble.

Controlling the forces that act on our torso is of fundamental importance. Having a weak musculature is thought to reduce our efficiency. Ideally, we need to be able to transfer the forces and maximize the capacity to use available

Figure 1.7. Daniel Scali holding the plank position for a world record 9 hr, 30 min, 1 sec.

potential and kinetic energy generated by everyday movements. One of the aims of this training program is to enhance the "control of systems that allow load to be transferred and movements to be smooth and effortless" (Lee, 2010). It is a sense of the quality of graceful, efficient movement, which is key to improving one's movement. This is quite different from a purely quantitative measure, such as holding a solid plank position for a set time. Learning to hold a rigid, solid plank position for 9 hr, 30 min, 1 sec, as Daniel Scali did during his world record, is a truly amazing achievement (figure 1.7).

However, this it is not necessarily an applicable measure of a functional core. My interpretation of a functional core is one of adjustability, adaptability, and fluidity. A functional core can manage, transfer, and use the forces generated by and through this area of our body. A functional core maintains a balance between mobility, strength, and stability. Too mobile, too strong, or too stable and the whole system will become unbalanced, increasing the likelihood of problems occurring. It is common that individuals in pain have muscular imbalances (Ellenbecker et al., 2009). Issues surrounding hypermobility are now well documented (Hakim et al., 2010), whereas a highly stable system is comparable with the stiff, immobile people I see daily in my clinic.

Core Language

If "language can corrupt thought" as George Orwell said, then language also seems to be able to corrupt action. Take, for example, the following statement from a well-known study:

> *The spine devoid of muscle and relying on its osteoligamentous framework is an inherently unstable structure and buckles very quickly under compressive loading. (Bergmark, 1989)*

As I read statements such as these, my thoughts may be "corrupted" into believing that my spine is unstable, vulnerable, and fragile. Such a catastrophic belief will likely affect the way I move as we know that movements reflect one's belief. Movements become extremely tentative; I am now fearful that my spine will buckle at any moment. Research tells us that flexion and extension, normal movements of the spine, are mechanisms for disc herniation (McGill, 2001). And so, I rarely, if ever, flex or extend and try to stay neutral and safe. I held that movement pattern of neutrality for nine years until more research came out. Nine years later, I read Veres et al., 2010, which states that a neutral spinal position is not an absolutely protective position to prevent disc damage. Now I'm confused, and as we know, confusion often heightens stress and worry. I find myself moving less and less, and so become increasingly stiff and immobile. This stiffness leads to pain to further support my belief of having an unstable spine. The vicious cycle of belief and action is complete.

Most of my working life has been as a clinician of bodywork. Within my clinic I regularly hear clients explaining that they have an "unstable spine" that needs "core support." Upon further investigation, I rarely find this to be true. Most often, in my experience, the ligamentous bed is intact, and the spine is functioning well. More commonly, muscles are overworking, gripping tight in a misplaced fear of overprotection.

To demonstrate the effects of overworking, try making a fist and holding it clenched for a minute, then an hour, then two hours. The result is fatigue, muscle cramps, and weakness. Imagine holding this fist clenched all day or all week and then suddenly adding a load, such as the dog pulling on its lead. These poor, tired muscles are unlikely to be able to respond quickly or efficiently to the dog. The core muscles are no different: if you are reading this with your core "engaged," tensed, or clenched, stop.

Let your breath descend into your belly to massage some fluidity back into your organs. Relax, and let your belly go, rest and be ready for the exercises later in this book.

Unfortunately, medical literature consistently suggests our spines are fragile structures. However, there is a growing realization that our vulnerable, weak, unstable spines have evolved perfectly well. Like many, I believe our spines are naturally flexible, strong, and resilient to the trials of sitting at the computer, running for a bus, or lifting heavy weights. It seems more amazing that the majority of people's bodies are not in constant pain and dysfunction, especially given the abuse we throw at them. The fact that of the 364 joints of the spine, only one or two may occasionally become uncomfortable is testament to the incredible resilience of our spine.

The idea of an inherently unstable spine is based on a strange reality—a reality where the muscles of the spine do not exist—"the spine devoid of muscle" (Bergmark, 1989). This notion of reality simply does not exist. The muscles are as integral to the spine as are the bones and ligaments. Anatomy, the segmentalizing or fragmentizing of the body, while useful theoretically, is often difficult to use practically in the understanding of the whole system in movement. At worst, this fragmented view creates poor, fearful movement patterns and cycles of belief that are difficult to change. Unfortunately, this fear-laden language is endemic within much of the anatomical literature. This catastrophic language has spilled out from research into the worlds of bodywork (physiotherapy, osteopathy, structural integration, massage, etc.), and movement (yoga, Pilates, strength and conditioning, martial arts, dance, etc.)

Various problems can be caused by our misuse of words. Our cognitive belief of a problem can create issues of fear, movement avoidance, and poor motor actions that can contribute to painful symptoms. As we now know, the perception of the threat of pain can be a cause of pain itself (Hodges et al., 2013).

Buddha reminded us to choose words with care as people will hear and be influenced by them, "for good or ill". Two words that are synonymous with core are strength and stability. Stability has become so bound to the word core that it is hard to mention core without the suffix "stability." Similar to the problems of defining core, "the term core stability has no clear definition" (Wirth et al., 2017). Throughout much of the literature, the terms stability and strength are used interchangeably and "Many studies fail to distinguish between core stability and core strength, two components that are fundamentally very different" (Hibbs et al., 2008).

Faries and Greenwood (2007) provide a useful explanation for the difference between core stability and core strength. They suggest that core stability refers to the ability to stabilize the spine as a result of muscle activity, with core strength referring to the "ability of the musculature to then produce force through contractile forces and intra-abdominal pressure (IAP)."

The language of stability and instability have strong connotations. Imagine meeting a stable individual. The character you meet is clearly grounded and balanced, a solid citizen, both rational and secure. Now you meet their cousin, who is rather an unstable person, muddled, unsteady, and erratic.

If you take on the personality characteristics of stability, you are firm, solid, and rigid. Alternatively, if you take on the features of an unstable personality, you are fluid, adaptable, and uncertain. Each characteristic portrays contrasting qualities of movement, laden with morality and expectation. Part of the definition of life, in Merriam-Webster dictionary, is an ability to "respond to change" (*Merriam-Webster Online Dictionary*, 2022).

From this we realize the need to adapt, like Lao Tzu's reeds, and respond to the demands of our uncertain world. It seems logical that a healthy core can respond to change and embrace its natural, inherent instability. As we know, within us lies our fluid visceral organs that need movement for hydration and health. Embracing uncertainty leads us to our complex (not complicated) and varied training. I hope this takes us away from ideas of a solid core, away from images of planks of wood, unmoving architectural structures, and machines that rust and stagnate, unable to change.

If "your body is a temple" (1 Corinthians 6:19–20), then architecturally, your spine would be a compressive, block-like structure. A temple is built block by compressive block, the bottom block dealing with the most load. If this was true for your spine, then each spinal block (vertebrae) would be balanced on the next, like the brick wall in your home. If this was true of the spine, then to withstand the compressive loads of standing, our lumbars would have to be 1.5 to 2 times larger than they actually are. If you increase the forces, the lumbars would somehow need to become bigger during walking, walking with a bag, running, weighted squats, and so on. Of course, this is impossible—the lumbars are not larger than they are—they are, in fact, the size they are! The logical conclusion is that the spine is not a compressive structure but one more akin to the ideas of tensegrity.

Core Balloons, Mechanics, and Tensegrity

Tensegrity, a contraction of the words *tension* and *integrity*, was conceived by Buckminster Fuller. Architect, systems theorist, inventor, and futurist, Fuller was one of the most fascinating minds

Figure 1.8. Kenneth Snelson's Needle Tower, *disappearing into the sky of Washington D.C.*

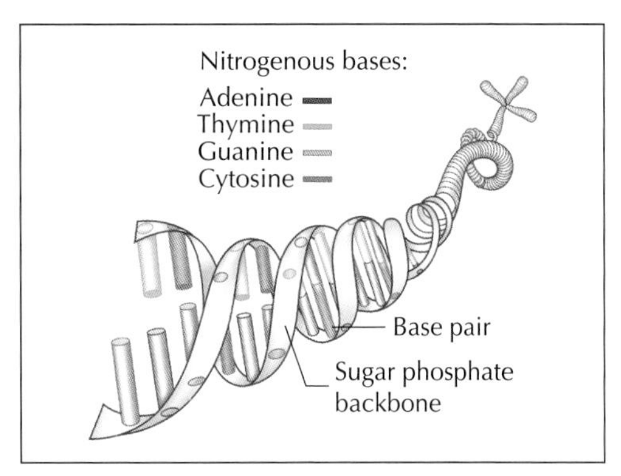

Figure 1.9. A simple representation of the building blocks of life, deoxyribonucleic acid (DNA).

of modern times. Arguably his most famous contribution was the geodesic dome. His student, the sculptor Kenneth Snelson, first created the tensegrity sculptures as expressions of nature's patterns of physical forces. Immediately recognizable, the struts of a tensegrity sculpture seem to float within a network of almost invisible cables. Combining the mechanical forces of tension and compression, the concept can seem rather esoteric. Snelson's *Needle Tower* soars to around 20 m (60 ft) into the sky (figure 1.8).

The sculpture continues until it touches the earth, and so addressing something basic both artistically and human. The structure reminds me of Franklin, Crick, and Watson's model of DNA (figure 1.9).

Both structures represent a fractal-like arrangement infinitely extending into the ether. Light, airy, and insubstantial, Snelson's tower moves in the wind, a direct contrast to the pillars of traditional stone sculpture (figure 1.10).

Others have described such tensegrity structures as analogous to a spider's web, yet neither the web nor the *Needle Tower* are true examples of tensegrity. At the heart of the definition of tensegrity is a

Figure 1.10. The Caryatid porch of the Erechtheion in Athens, Greece, showing the architectural support of a compressive structure.

shape that is "finitely closed" (Fuller & Applewhite, 1982). This means a true tensegrity structure is not anchored to a solid structure like a spider's web or the *Needle Tower* must be. The tensegrity structure is independent and can maintain its basic shape in any position or environment.

A simple tensegrity model (figure 1.11) closely resembles a box that is akin to our own torso. The simplest of tensegrity structures has numerous characteristics that make it a highly useful tool for the explanation of the core and indeed the whole body.

> *Explaining the motion, interconnection, responsiveness and strain patterning of the body without tensegrity is simply incomplete. (Myers, 2020)*

One of the key characteristics of a tensegrity structure is that it distributes stress and strain. Drop a tensegrity structure (e.g., figure 1.11) and it will morph and distribute the forces as it hits the ground. The elastic guywires of this tensegrity structure allow it to bounce back into shape. Compare this to when I dropped my phone: with a sickening "thud," it hit the tiled kitchen floor. The impact of the force was concentrated to crack the screen and break the whole back into its many component parts. In contrast, the tensegrity structure, like the body, distributes strain, dissipating it across the system. If you jump, the impact force is distributed around your body rather than being concentrated into one, probably painful, area. This goes someway to explain why twisting your ankle could result in back pain later in life. The initial injury creates a reduction in the ability to distribute force, resulting in more force concentrated into one unhappy structure.

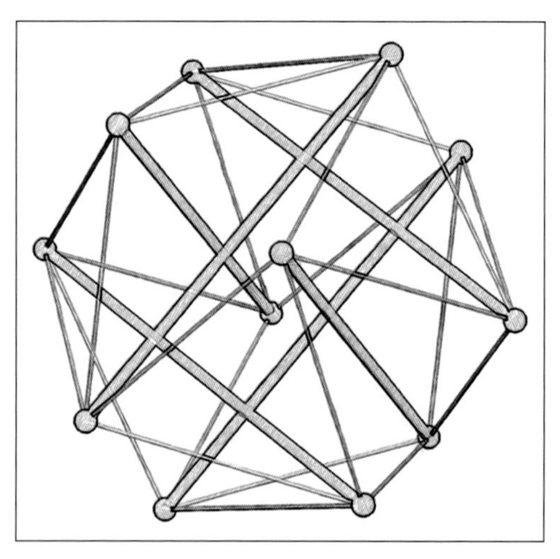

Figure 1.11. A simple tensegrity model. Image from Movement Integration: The Systemic Approach to Human Movement, *Lundgren & Johansson, Lotus Publishing, 2019.*

While this tensegrity model is a useful one to describe the interplay of hard and soft tissues, it only tells part of the story.

The core is often described as a box "with the abdominals in the front, paraspinals and gluteals as the back, the diaphragm as the roof and the pelvic floor and hip girdle musculature as the bottom" (Akuthota et al., 2008). This multi-walled thermos flask acts and reacts to numerous directions of force in the transfer of load and distribution of energy. The analogy of a box or cylindrical thermos flask goes some way to describe the three-dimensional nature of our core. It also conjures up unhelpful images of a rigid form.

Our torso resembles a more mobile tensegrity-like structure: a balloon. The IAP (the air in the balloon) acts to tension the myofascia (the silicon fabric of the balloon) creating shape, resilience, and stiffness to the structure. The fabric of the balloon, like the myofascia, has elastic and plastic properties.

Unlike the balloon, however, our own fabric is not a single sheet of fabric and has the capacity to adjust its relative stiffness depending on the demands. This stiffness-adjusting capacity is a characteristic of biotensegrity and highly important for a functional core.

To enable this balloon to better resemble our torso, let us add a little more sophistication to this analogy. Imagine that our balloon is made not of a single silicon skin but of bubble-wrap (blister pack). Note how each bubble is continuous with the whole plastic structure of our strange balloon. Now imagine that each bubble could expand, fill with more air, and enlarge. This expansion would increase the overall surface tension and increase the stiffness of the balloon. If we let the air out of a few of the bubbles, and the surface tension of the balloon is reduced, our bubble-wrap balloon would become softer and more pliable. Our muscles are analogous to these bubbles; a contracting muscle expands and swells outward—pulling on the fascial tissue—increasing the relative tension of our abdominal balloon. In this way, the tensioning of one muscle communicates to other muscles to respond and adjusts the stiffness of the whole structure. This process of "hydraulic amplification" is thought to increases the muscle efficiency by around 30% (Earls, 2020).

Hydraulic amplification is not confined to the torso but can be found throughout the body. The expanding quadriceps muscles of the thigh, for example, push out into the fascia lata of the leg increasing muscle efficiency during a squat. If the balloon bursts or is cut, then the efficiency of the muscles is compromised. During a fasciotomy (a surgical procedure for compartment syndrome), a thin incision is made into the fascial sock of the lower leg reducing the muscles's efficiency by 12%–16% (Gracovetsky, 1988).

Our abdominal balloon has the amazing capacity to be both mobile and stable at different times depending on the requirements. Attempt to lift an Atlas stone, which typically weighs anything from 242 to 352 lb (110 to 160 kg), and you'll certainly need a stable core. During such a lift we need the capacity to stiffen and make our fluid core solid. Lift anything heavy and you'll soon

Figure 1.12. Weighted Squat.

appreciate that the power mostly comes from the legs, which is transferred though our torso to our arms. Weighted squats (figure 1.12) encourage the core to take on a stiffness strategy for success in load transfer. During these high-load, predictable situations, one can pre-tension, create stiffness, take, and hold a breath, and lift exceptionally heavy weights effectively.

This strategy of pre-tensioning and contracting the core muscles compresses the viscera and increases the pressure in the abdominal balloon. This IAP is greater if the breath is held after a deep inspiration. Weightlifters often use a breath-hold strategy, occasionally turning purple as they lift. This technique uses the breath to lower the diaphragm and so reduces the relative size of the abdominal cavity. The pressure increases to make use of the hydraulic amplification for muscle efficiency. When the torso is made into a relatively solid cylinder by axial compression, shear loads are reduced and transmitted or dispersed over a wider area. Then as one lifts or moves, the expansion of the muscles acts to further increase the capacity of the abdominal balloon. When we contract a muscle within its fascial bag (balloon), the muscle expands, increasing its volume. The effect is like blowing more air into a balloon, increasing the surface tension, and so increasing the efficacy of muscles to transfer force across this changeable structure. The result is a core that is further stiffened as the work of the muscles is increased by the demands of the activity.

Where this simple bubble-wrap balloon analogy fails is that it suggests a linear relationship of muscles neatly lined up. The complex reality is that these muscles are arranged at different layers, depths, positions, and sizes. Some of this complexity can be seen in cross section (figure 0.1) and is due to the multifunctional, multidirectional role of all muscles, only one of which being hydraulic amplification.

Taking the above information in isolation, one might consider that a solid core is ideal. Indeed, this poor conclusion has often been made. A solid, braced core will, over time, reduce the integrity and efficiency of our much-needed elasticity. Constant loading and the muscles will fatigue, become weak, and stiffen. Such a strategy will compress the spine, giving greater stability in the short term but has the potential to cause more issues in the longer term.

Stability and mobility are so intertwined that Dr Gary Gray, the self-proclaimed "Father of Function," coined the word "mostability." Our core certainly needs a mobility strategy just as much as it needs a stability strategy. Unfortunately, the popularity of core stability has become such a "thing" that some have failed to realize the necessity for core mobility. The capacity to "let go" and deflate our balloon allows the mobility necessary for movement. Various styles of yoga use the breath to vary the IAP to allow the body to achieve greater ranges of movement. Ensuring we regularly let our belly go, sag, and relax is important to our health. Regular mobility work to allow length where there might otherwise be only compression allows the core better homeostatic balance.

Allocating time for mobility exercises, such as the psoas stretch (figure 3.35), allows the decompression of the spine after more maximal lifts. Decompressing-style mobility exercises are also significantly important after long intervals of sitting. Many desks and chairs seem to have been carefully designed to shorten the psoas, hip flexors, and squash our viscera. The mobility exercises in this text will go some way to alleviate this, although the best thing is simply getting up and regularly moving.

Exercises such as a weighted squats (figure 1.12) encourage a stiffness strategy of the core. During this high-load, predictable situation, one can pre-tension, create stiffness, take a breath, and lift. The strategy of breath control uses the effects of hydraulic amplification by using the IAP to pre-tension, stiffen, and so enhance force transfer during these predictable actions. But not all lifts are so predictable; lifting a squirming infant by using a pre-tensioning, breath-holding strategy is neither practical nor needed as they are not particularly heavy. Nor is it useful to be a fluid, flexible, relaxed holder of an infant, for they seem very heavy and, contrary to the popular myth, they don't bounce when dropped!

The strategy used for a simple squat is even less useful when walking the dog. The squat is predictable, whereas the pull on the dog lead is sudden and variable. In my opinion, training needs to include ways to educate our bodies to react quickly to mirror some of the uncertainties of life.

Instead, we need a combined approach that allows the core balloon to be "just right," somewhat stiff, and somewhat mobile; to be strong enough without being rigid, and yet supple enough to deal with the ever-changing force of everyday life. Exercises with this style can be found throughout this book. These exercises help facilitate a strategy using more neuromuscular control to educate the body during low-load, unpredictable situations that mirror some of the uncertainties of life.

CHAPTER 2

Delving Deeper

Society does not consist of individuals but expresses the sum of interrelations, the relations within which these individuals stand. (Karl Marx)

The Social Core

The intention of considering the core from a social context is to take the exercises from the gym into more real-life situations. Society has proved to be a force with the potential to create a sinister relationship with our own body. It is hoped that awareness and careful scrutiny will somewhat nullify this negative social force. Our style of movement, training, and exercise is not a result of muscles, ligaments, fascia, and bony anomalies alone, but it is driven by our beliefs, that are in turn driven by social norms, fashions, significant others, and the wider environment. The objective of considering the core from this wider context is to avoid the pitfalls of history and allow training to be more functional, practical, and useful.

Our beliefs as to what constitutes a "good core" is often echoed by our beliefs regarding "good posture." The common narrative is that we need a solid core to support our weak and vulnerable spines. So, sit up straight as you read this, brace your core, and you too can obtain perfect posture and the sun will always shine. Fail to comply with this common belief and be condemned to slouching into a life of back pain and rain!

Shockingly, this sentiment still prevails (perhaps without the weather report) and advertisers still use these outdated ideas to sell various core products and core exercise plans. Thankfully, good research coupled with common life experiences show that the link between posture and pain is nothing more than a myth (Dillner, 2018).

You may well know someone who you would describe as having good posture yet is in pain, and someone else who has appalling posture but is not in pain. This common experience is at odds with the belief that posture and pain are somehow correlated. This belief has deep roots that can be traced back though our history. Our historical habit of linking posture to pain has created more issues than it has helped. Posture and pain are worthy of separate and lengthy books, and I direct you to the reference page for some examples.

Posture and Pain

To truly laugh, you must be able to take your pain, and play with it. (Charlie Chaplin)

It is rare to find clarity and agreement within research. However, on the topic of pain and posture, there is increasing agreement that pain is not caused by posture and posture is not a cause of pain (Korakakis et al., 2019). The complexity of pain, as shown in the research, demonstrates that pain is a result of numerous multifactorial interactions.

Attempting to explain pain from a reductionist, cause-and-effect viewpoint (such as improve posture to improve pain) is like looking for a single "magic bullet" treatment. Traditionally it was posture that was used as the mythical holy grail of pain. I certainly do not wish to add to this mythology by suggesting that improving one's core will somehow be the next cure-all to pain.

A few moments on Google Scholar shows the significant amount of research that has attempted to correlate core strength and core stability to lower back pain. The inconclusive results remind us that the reason for lower back, or any, pain is specific to each individual and task. For one person, such exercises may help reduce back pain—however, for another person, they may not—each time it depends upon the underlying cause of the pain.

Understanding pain is to embrace the uncertain complexity of relationships. Pain is not a simple reaction to tissue damage because people may experience pain with no tissue damage (Moseley & Butler, 2013). Perhaps the most dramatic example of this is phantom-limb pain: the situation where someone is experiencing real pain in their hand but has no hand or even an arm (Moseley & Butler, 2013).

For most of us, a more everyday example occurs in the bath. As you take a bath you might notice a bruise or scratch on your leg. As your mind wanders back through the day, you realize that this tissue damage has no associated pain or even a memory of a painful event. (For more information on pain, I refer you to the pain sciences version of the Bible: *Explain Pain* (Moseley & Butler, 2013).

To further understand the complex relationship between posture and pain, let me tell a story from a client with whom I worked recently. The interaction between the old man, his wife, and myself went something like this:

Me: How can I help? My usual open-ended, hopefully friendly initial question that places the client's needs and goals at the heart of our interaction.

Wife: I want you to get him to stand up straight, he's always stooped forward. Spends more time looking at his feet than at me!

Me (to the wife): And why do you want him to stand up straight?

Wife: Well, that back of his hurts him (so often "the back" becomes a separate entity, almost like an enemy suggesting a difficult relationship).

Me (to the husband): When does your back hurt?

Husband: When I stand up straight.

I hope you see my dilemma. This brief interaction tells how this stooped-over old man held his so-called poor posture to avoid pain. The pain, we later discovered, was caused by spinal stenosis (a narrowing of the foramina within the spine). His poor posture was an effective way to avoid pain and function. Had I complied to his wife's wishes and tried to improve his posture toward a set of aesthetic social norms and expectations, I would likely have created more pain. Unfortunately, this story is all-too familiar in my clinic and for many other clinicians.

There continues to be no agreed definition of what constitutes good or bad posture.

Unfortunately, much of the history of bodywork was underpinned by the inaccurate belief that improving posture would reduce pain. It is unfortunate that some schools of bodywork still promote this limited idea. Worse news is that this restricted and restricting view is not confined to the world of bodywork. Numerous schools of exercise also promote good posture as a gold standard to achieve, often without a clear definition or understanding of the supposed benefits. Often, achieving good posture is nothing more than conforming to the fickle nature of current postural fashion.

The implication that a good posture exists is endemic in public opinion, research, and various schools of bodywork and movement. The term *good posture* suggests a singular, a one-size-fits-all, concrete position to somehow achieve. This perspective, I believe, places a limit on our understanding and movement possibilities. I take the view that posture is a far more complex shifting pattern of reflexes, habits, and adaptive responses to the demands of life. Good posture is a dynamic response to the interactions and accompanying feedback loops within our constantly changing body that acts and reacts as a consequence of our uncertain, nonlinear environment.

Research into posture consistently uses language that refers to "good" or "bad" posture. Exactly what is meant by good or bad is rarely defined, perhaps because the concept of good and bad is a theological—rather than anatomical—discussion. The consistent use of the terms good and bad, when referring to posture, indicates the strong connection of posture to our social morals.

In his book *Stand up Straight!*, Sander Gilman combs through our tangled relationship of morals and posture (Gilman, 2018). Gilman explains how the morality of posture has been with us since at least 8 CE when Ovid wrote *Metamorphoses*. *Metamorphoses* has influenced humanity's expression of self through its art, inspiring the likes of Dante, Chaucer, and Shakespeare. Within these pages, Ovid observes that while "all other animals are prone, and fix their gaze upon the earth, he gave to Man an uplifted face and bade him stand erect and turn his eyes to heaven" (Ovid, 800).

Figure 2.1. Gru, the hunched supervillain from Despicable Me *(Universal Pictures).*

Continuing this train of thought, one wonders what might become of the slouching teenager, a posture perceived to demonstrate an inferior psychology—devoid of the divine—rather than a stage in development that most of us physically and posturally mature out of.

It is a common technique in literature to describe the posture of a character to demonstrate their personality. From the fierce flirtations of Bizet's Carmen with hands on hips, eyes over shoulder, or Gru, the hunched supervillain of *Despicable Me* (figure 2.1).

Regrettably, in my opinion, this literary trick of using posture as a window into the soul of a character has leaked off the page into clinical reality and everyday belief.

For many, postures have become the psychological made physical. The common practice of reading one's posture seems similar to phrenology—the Victorian pseudoscience that suggested that the bumps on one's head could explain personality traits and character. Taken to its extreme, phrenology was used to justify slavery and racial inequality.

The suggestion that a particular posture is representative of a corrupt moral state

Figure 2.2. David by Michelangelo in the V&A, London.

has significant implications. If this idea was confined to the artistic world of plays and poems, I would not mention it—but as is so often the case—art reflects and is reflected by its social context.

I love to spend my time studying the idealized forms of classical sculpture in the V&A, London (figure 2.2). Part of my fascination is identifying the occasional anatomical errors of some of these sculptures. These errors were likely created, not because of a lack of anatomical understanding, but as a method to enhance the aesthetic of these godlike bodies. Such sculptures are often aspirational representations of the ideal, godlike form; a body designed by God, "the supreme architect" (Giovanni Pico Della Mirandola, 1496). Underpinning such pieces of art is the notion that we all have the potential for perfection that reflects "His own image" (Genesis).

If sculpture is the looking glass of society's perception, then our corresponding postural expectations are perhaps changing. This idea is reflected by the popularity of a new style of sculpture typified by Anthony Gormley. Gormley includes his own physical idiosyncrasies into his sculpture. His sculptures mirror humanity's uniqueness through the medial tilt of his left calcaneus, his imbalanced shoulder girdle, translated fifth rib (I could go on!) that can be seen in many of Gormley's works. Postural imbalances that Michelangelo surely would never have allowed in his representation of David.

The renowned pianist Vladimir Horowitz who, when asked about his seemingly perfect playing, explained that "perfection itself is imperfection" (Plaskin, 1984).

So it is that moving away from the ideals of a perfect posture—as represented by ancient godlike solid stone statues—allows us to embrace a postural reality of fluidity, uncertainty, and imperfection. A posture reflecting the mobility of Snelson's tensegrity and Gormley's flawed humanity.

Core Fashion and Corsets

I see that the fashion wears out more apparel than the man. (Shakespeare)

Fashion is a form of ugliness so intolerable that we have to alter it every six months. (Oscar Wilde)

In our attempts to achieve godly posture—and with it, moral perfection—society has inadvertently created numerous methods of psychological, social, and physical abuse, pain, and ill health. One such example lies in the clothes we wear. Model Joan Smalls reminded us that fashion is more than just a pretty dress—it is part of our culture.

Before the mini skirt, which kept knees together during the sexual revolution, came perhaps the most controversial—and for us relevant—article of clothing: the corset. The term *corset* is an overused analogy used to describe the muscular core. An appreciation of the corset will, I hope, go some way in preventing the core becoming the new corset. A garment beset with physical and psychological hazards and problems.

In her book, *The Corset: A Cultural History*, Valerie Steele explains the complex relationship we all have with this item of clothing. Today the corset has become almost universally condemned as a tool in the oppression, control, and sexual exploitation of women. The aim of corsets was to

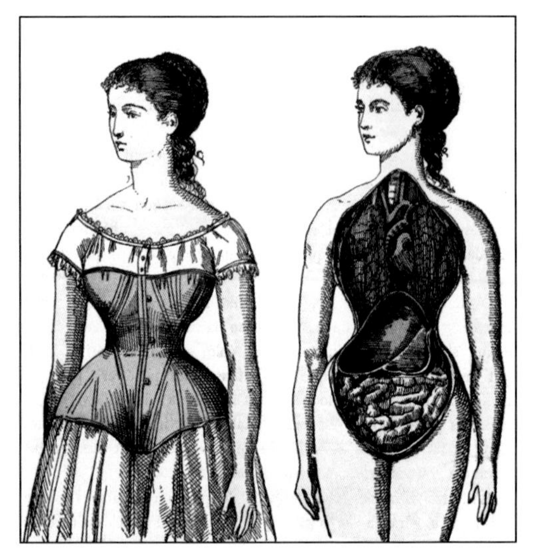

Figure 2.3. Representation of how the corset might affect the ribs and organs.

attempt to conform the wearer to have an idealized wasp-like waist with crushed ribs and internal organs. Numerous reports tell of a range of ailments (not all true) from fainting, to cancer, hernias, and even death that led *Punch* magazine to call the corset "fashionable suicide" (figure 2.3) (Steele, 2003).

Men made the stays or supports of early corsets from whalebone, horn, and occasionally metal. French army surgeon Ambroise Paré designed metal stays to "amend the crookedness of the body." In 1880, *The Lancet* reported that prolonged wearing of a corset resulted in atrophy of the musculature. Paradoxically this weakened muscular state led to the need for more support. The fact that devices—like Paré's metal corsets—are still available today, perhaps indicates how little we still understand of our bodies in movement.

This vicious cycle of support and consequential weakness is not confined to the corset. Office chairs are "specially designed to enhance comfort, posture, and wellbeing" (from a well-known chair manufacturer). Children's shoes are sold with the promise to support supposedly weak arches. At first this support is unnecessary, assuming the muscles haven't yet atrophied. With time, and the ensuing reduction of strength and resilience necessary, the support offered becomes a habitual dependency that is hard to quit.

The demise of the corset was not a result of the mountain of medical ailments associated with it. Instead, it was exercise that saved us from the corset. The women's rights movement—and its sister, dress reform—encouraged women to exercise, be free to move, and have a love for the outdoors. One result of this reform was the rapid demise of the solid corset and all it represented.

Unfortunately, the detrimental effects of the corset can still be seen today in the exercise ideals of having a solid, stable core. In certain situations, you may be told to pull your belly into your spine, to lace-up, tighten, and brace your core. This trains the core to act in much the same dysfunctional way as the early corsets. While unlikely to cause death when laced tight, our "corset-like core" can—like a solid corset—alter rib positions and affect breathing patterns.

The sad result of such a rigid idea of the core has, for many, become the internalization of the corset through diet, exercise, and even plastic surgery. The consequence can be mental illnesses, eating disorders, and the creation of a sinister relationship with one's body image. The idea of a solid, perfect, and certain core is a concept that undermines freedom of movement, equality, and individual differences. It is hoped that these pages will offer an alternative notion to core training and—with this awareness—avoid the pitfalls taught to us by the corset-wearing history teacher.

Function and the Military

It is somewhat ironic that much of our understating of functional movement comes from the military. The military world evokes conformity to standing correctly at attention. The absorbingly insightful and, at the time, controversial work of the photographer, Muybridge attempted to analyze the optimal fighting posture (figure 2.4)

Most famous for his motion-study photography of galloping horses, Muybridge used his insights and ingenuity in the analysis of human movement. Chin up, chest out, stomach in, shoulders back is a classic military posture that evokes a solidity of standing to attention. However, the ideal fighter is more mobile than solid. The mobility and uncertain tactics of David were in direct contrast to the solidity of Goliath who wore a "helmet of brass upon his head, and he was armed with a coat of

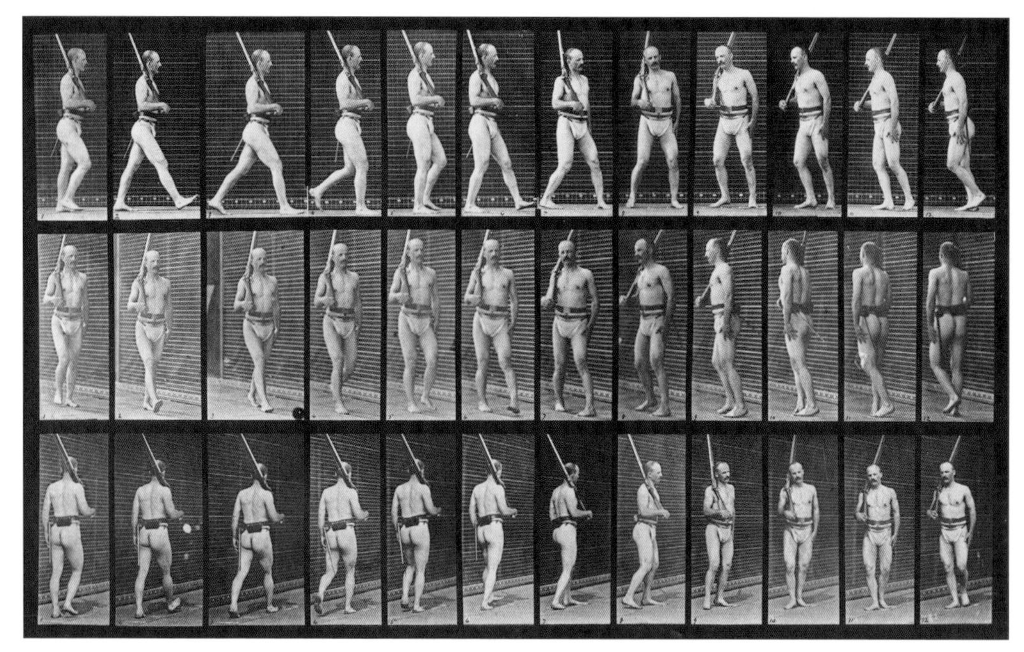

Figure 2.4. Eadweard Muybridge, 1887, Man Pacing With Rifle.

mail; and the weight of the coat was five thousand shekels of brass. And he had greaves of brass upon his legs, and a target of brass between his shoulders" (1 Samuel 17:5–7).

Figure 2.5. Rock Drill *by Jacob Epstein (original work dated 1913–1916).*

Renowned for refusing to be inducted into the US army, boxer Muhammad Ali did not have a solid military posture. Perhaps this emphasis on movement and adaptability over solidity and stiffness was the inspiration for Muhammad Ali, who famously "float[s]like a butterfly and sting[s] like a bee" (Remnick, 1999).

The military rigid posture seems to dehumanize the individual into a metallic military weapon that resembles Jacob Epstein's infamous *Rock Drill* sculpture (figure 2.5)

For some—like Forest Gump—the stereotypical unquestioning certainty of the army seems to fit. "Now for some reason, I fit in the army like one of them round pegs. It's not really hard. You just make your bed real neat and remember to stand up straight and always answer every question with 'Yes, drill sergeant.'" (Zemeckis, 1994). For others, conformity to a correct rigidity is not so easy. In 1962 the legendary musician Jimi Hendrix was honorably discharged from the US army with a report describing Hendrix as an "individual who is unable to conform to military rules and regulations" perhaps suggestive of his creative musical brilliance that lay far outside convention (Henderson, 2008).

Figure 2.6. Guard at Buckingham Palace, London.

Attempt to stand for a prolonged period of time, and—like the statuesque guards at Buckingham Palace (figure 2.6)—you'll soon realize that even the most static posture is in a state of continual flux. Listen and your sensitive feet (with help from over 7,000 nerve endings) will tell you of their subtle changing patterns of work—a shifting flow of adjustments that creates our natural postural sway.

This sway has useful economic benefits (in terms of energy) to ensure we remain efficient. If we really could stand as still as a statue, parts of our body would soon become overburdened by such constant work, resulting in chronic fatigue, weakness, and eventually failure. Instead, our system monitors the relative workloads, distributing and changing the type and amount of work accordingly, like an orchestral interplay accompanying the interacting duo of gravity and ground through the instruments of proprioceptors, exteroceptors, muscle spindles, Golgi tendon organs, vestibular oculomotor system, and muscle reflexes (to name just a few of the complex interacting systems that allow the dynamic posture of standing "still").

Humans are amazing generalists, not the specialists we find elsewhere in the natural world. We can reach great heights and explore the ocean depths, yet struggle to reach the tops of tall trees with the ease of a giraffe or swim with a sailfish at 70 mph.

Humans struggle to be specialists—repetitive, specialist work soon becomes monotonous as well as physically and mentally boring. If humans attempt to act like robots—duplicating repetitive movements—we soon create the potential for overloading and overworking our bodies. The irony for me is that writing these words is an example of a repetitive dysfunctional action. In an attempt to mitigate the harm associated with sitting at my desk, I take movement breaks every 20 minutes. I also vary where I write and sometimes sit, stand, or spend time on the floor. Variety of position changes the load distribution and workload within our system, and so changing the nature of our daily work can be highly beneficial.

I'm lucky that working in my clinic allows an active lifestyle to offset the moments of sedentary writing. However, many jobs and lifestyles are so specialized that they do not allow for the variety that the body and mind crave, which is where varied exercise programs become essential.

A few years ago, I walked the Pennine Way, 268 miles along the spine of the UK. During each day of walking, I would take a few breaks and sit. I did not take a break by walking—taking a walking break during a day of walking sounds preposterous—yet sitting after a day of sitting is common. A familiar pattern to daily life is to drive to work, sit at a desk, then drive to the gym only to sit on a bike, or a weights machine, then drive home and rest by sitting to watch TV. Interspersing this daily pattern with variations from sitting will give an improved and more varied "diet" of movement with its many benefits. A simple yet effective suggestion is to take a couple of short walks during a day that is predominated by sitting.

The problems associated with repetition is not confined to work or long walking holidays. When exercising, it is common to repeat the same actions and the same exercises. Often such habitual training is born from specific goals of success. Perhaps you learn the plank exercise, enjoy the action, and achieve a measurable goal of planking for a specific time. This might lead to setting new goals perhaps, using the powerful motivator of **SMARTER**—**S**pecific, **M**easurable, **A**ttainable, **R**elevant, **T**imed, **E**valuated, **R**eviewed—goals. While goal setting is a brilliant way to improve our motivation to train, it can result in repeating the same habitual movement pattern. To some extent, this does help achieve such measurable goals. However, few of us really desire to be a world record "planker" like Danial Scali (figure 1.7). I am still in awe as to how Scali held the plank position for 9 h, 30 min. He demonstrated that some of us can become highly specialized. However, the vast majority of us are naturally generalists.

Varied training allows us to be better prepared for the variety of possible daily challenges.

> *[A] breadth of training predicts breadth of transfer. That is, the more contexts in which something is learned, the more the learner creates abstract models, and the*

less they rely on any particular example. Learners become better at applying their knowledge to a situation they've never seen before, which is the essence of creativity. (Epstein, 2020)

This dynamic understanding of posture, and the part the core plays in our complex body, mirrors a major repeating theme across modern science. The sciences consistently tell us that all natural phenomena are ultimately interconnected and their essential properties derive from their relationship to other things.

Quantum theory reveals a basic oneness of the universe. It shows that we cannot decompose the world into independently existing smallest units. As we penetrate matter, nature does not show us any isolation "building blocks," but rather appears as a complicated web of relations between various parts of the whole. (Capra, 1992)

This Zen-like appreciation of the nature of physics also means our core structures cannot be separated, trained in isolation, and somehow reinserted to improve our body, our posture, or our mood. Instead, the exercises advocated in these pages need to be part of the larger whole of training our body, developing our mind, and improving our environment. The aim is to avoid the physical entrenchment of repetition and to develop an adaptable, varied, curiously creative system that is able to improvise.

History of Division

There is more power in unity than division. (Emanual Cleaver)

I hesitate to list the muscles associated with the torso lest they somehow become perceived as separate entities. Analysis, anatomizing, and dissection finds division where there is none. The scalpel's capacity to create isolated muscles have, for some, anthropomorphized individual muscles into having their own character and hierarchical sense of grandeur and importance. In this way, muscles have become characters in a play, a favorite muscle becomes the hero to worship and study.

In Daniel Kahneman's explanation, the agent is the muscle and "the mind … appears to have a special aptitude for the construction and interpretation of stories about active agents, who have personalities, habits, and abilities" (Daniel Kahneman, 2012). Individual muscles with higher level importance do not exist. The reality is that no such individual structure or hierarchical organization of importance exists for any organic structure. The stem of a tulip is no more or less important than its flower. The TrA is no more or less important that the internal obliques. Often the unequal weighting given to the research of one muscle, such as the TrA over another, suggests a level of importance than does not reflect reality. Our system is not hierarchical; rather a heterarchical organization exists.

Within a heterarchical organization, no one element claims dominance over the rest. Instead, it is circumstance that drives the organism's organization and reorganization in a complex display of interdependency.

Anatomical knowledge of the body is still largely based upon hierarchy and division. At the top of this hierarchy was—for many years—the scalp-wielding Roman anatomist Galen (129–216 CE). For thousands of years, Galen's anatomy taught us a lesson in medical certainty. His treatments worked, he claimed, on everyone except those who were going to die anyway—a claim hard to argue against and showed an arrogance—once disemboweling a live monkey and challenging the physicians in attendance to replace its organs correctly (Mattern, 2013). Galen's anatomy became a gospel; anatomists acted like monks, devotedly copying this sacred anatomy bible.

Anatomists of this time interpreted their findings according to their prior theoretical assumptions based on the work of Galen. Incredibly skillful as Galen certainly was, he did make mistakes, not in itself a problem as "mistakes are the portals of discovery" (Joyce, 1914). However, mistakes that are repeated unquestionably fail to open the portal of discovery.

The anatomy world had to wait until the questions of Leonardo da Vinci (1452–1519) and subsequent achievements of Vesalius (1514–1564). Many of Galen's errors were a result of only being allowed to perform animal dissection, which he then theorized would be the same as when performed on humans. Galen's anatomy had the human mandible as two bones—like a dog—rather than one. Ironically, Galen's animal-based theories allowed humans to be perceived as separate from—and have a higher status than—the animals of his dissections.

These theories were powerfully supported by the church, perhaps explaining their longevity. In one of the longest examples of the self-fulfilling prophecy, the church defended Galen's theories that neatly fitted their own beliefs. Galen believed that the brain generated and transmitted a vital spirit through the hollow nerves to the muscles, allowing movement and sensation (Mattern, 2013). Today, Galen teaches us a useful lesson of certainty and social pressure. The combination of church and anatomist had the power to create a perception of the body that was, at times, just wrong.

Approximately 1,700 years after Galen's death, it was Vesalius who became known as the founder of modern human anatomy. Vesalius took a more skeptical approach to Galen's work. It was through empirical observation that Vesalius shaped his understanding of human anatomy. Vesalius conformed to the reality of the external world rather than being unduly influenced by existing assumptions. By questioning the work of Galen, Vesalius taught us a lesson of uncertainty. This idea of questioning everything is now the bedrock of modern-day science.

> *History teaches us that man learns nothing from history. (Hegel)*

History teaches that attempting to place people into particular categories is difficult and potentially dangerous. Attempting to categorize humans into neat little boxes has led to the horrific consequences of racism, sexism, and antisemitism.

It is worrying to observe the parallels in the attempts made to categorize our own bodies. One such example of this is the work of the psychologist William Sheldon. Sheldon was not alone nor the first to be convinced that there was a link between physique, personality, delinquency, and criminal behavior. The theory—*somatotyping*—placed people into one of three categories based on their physical shape. These three physical morphologies were attributed to personality tendencies to further categorize each person.

Seen as revolutionary in the 1940s, the theory captured popular imagination. The theory of somatotyping is now discredited by the scientific community. However, it is still used in gyms today as a measure of body type, despite its lack of evidence and the fact that it was devised by a psychologist not a physiologist.

The concern with the continued acceptance of somatotyping by some is its links with the concept of eugenics. The theory of eugenics states that people with good genetics should procreate with each other, and people with bad genetics shouldn't procreate at all. Eugenics was a theory at the center of the Nazi party. Sheldon's brand of eugenic race science lead to the idea that it was "inbreeding that caused Jewish poor posture and bad character" (Gilman, 2018). Sheldon attempted to distance himself from these links after the Second World War, but the damage had been done and taken to its horrific conclusion.

However hard the scientific community has tried, accurate categorization remains difficult and elusive. Bodies, it seems, like people defy categorization, perhaps because of the multiple roles and fickle ability to change the characteristics depending on the conditions.

For a few years I worked as a teacher, educating students how to pass exams. To keep my job, I kept to the national curriculum and perpetuated the idea that muscle fibers could be placed into two unimaginatively named categories: type 1 or type 2. The idea was that type 1 or slow twitch fibers were muscles resistant to fatigue, work at low load, and their primary job was one of postural support against gravity. This was contrasted by type 2 or fast twitch muscles, which fatigue quickly and provide shorter more powerful forces.

However neat this categorization seems, it is now considered to be flawed, not exactly wrong but certainly an incomplete and simplified explanation. Research shows us that during fast running, the TrA and the diaphragm, traditionally placed in the slow, type 1 category, show faster, type 2 characteristics (Saunders et al., 2004). Another example of the complex nature of our system defying categorization.

The problems associated with categorization continue with the attempt to divide muscles into either local or global muscles. It was thought that the longer global muscles, which often cross two of more joints, are better suited for the role of creating motion. And so, the location of the shorter local muscles acts to create stability, often affecting one specific joint. A neat theory that reality calls into question. Muscles, whatever their category, can and do take on roles of both stability and mobility. In fact, they can switch roles during an activity as the needs arise. It is often the case that theories of the past contain elements of truth, even if the overall concept is now dismissed. There is truth in that shorter muscles, like the multifidus, are well positioned to stabilize joints and that longer muscles, like the rectus abdominis, are better positioned, due to their longer moment arm, to create movement.

However, we now realize that these long global muscles can stabilize the torso by exerting a compressive force. This strategy has been found to be effective in the short term although not as a long-term strategy. To win an Olympic medal, weightlifters sometimes compromise their system and use the capacity for muscles to switch roles in order to generate and manage the massive forces during a lift. It is clear that long-term compressive forces are detrimental to the health and motion of the spine (Pope et al., 2002). Holding long-term tension in global muscles can contribute to other issues. For example, using the external obliques to maintain spinal stability can negatively affect breathing patterns. A lack of breath directed into the lower and lateral part of the lungs reduces the capacity of the lower larger lobes of the lungs and may be a result of excessive use of the external oblique for core stability.

The belief that smaller local muscles are involved primarily with core stability, whereas the larger global muscles are involved primarily with force production "... has prompted ineffective training strategies designed to train the local and global muscle groups separately in nonfunctional positions. For example, the abdominal draw-in maneuver, in a quadruped or supine body position ... These muscles function as kinesiological monitors that provide the neural subsystem with proprioceptive feedback to facilitate coactivation of the global muscles to meet stability requirements." (Willardson, 2007)

It is clearly time to move away from such division and categorization toward a more synergistic method of training movements, not training specific muscles or groups of muscles.

Identifying our core has, so far, led us into an inconsistent world of fluidity and flux. The concept of our core as our center has no consistent location and so defies consistent definition. This truth was discovered by Archimedes in the third century BCE: "our centre has no consistent location" (Archimedes, 287 BCE). Archimedes realized that the center of mass of an object with variable shape, such as the human body, changes. The precise position of our center will change with the position of our limbs and the positions we happen to be in. Our center sometimes falls inside, sometimes outside of our body. These changes challenge our resilience, balance, and adaptability to manage each change of situation and demand. This idea gives more support to the idea that the core is a changeable, adaptable concept.

CHAPTER 3

Muscles

Core Flag

The structural fabric of the torso has a warp and weft that displays the effects of the forces of gravity, ground reaction force, mass, and momentum. As D'Arcy Thompson states, the form of an object is a "diagram of forces" (Thompson, 1917). This is like how the rings on a pot show the centrifugal forces: a pattern of forces that tells of the change from a lump of clay to a functional art form (figure 3.1).

Figure 3.1. The spiraling rings of this magnificent pot tells of the forces of its creation.

Our lines of force are organized much like the flag of Great Britain and Northern Ireland, nicknamed the Union Jack (figure 3.2a). I first heard the idea of a flag of muscles when in conversation with author, lecturer, and bodyworker, James Earls, an idea that can be discovered in his book *Fascial Release for Structural Balance* (Earls & Myers, 2017).

The Union Jack flag was designed by another James: James I in 1606. The flag combines three older flags to show the collective union of nations: The English cross of St. George, the Sottish white saltire of St. Andrew, and the Irish red saltire of St. Patrick. This combined approach to flag design mirrors the core concepts of multiple aspects conjoined to create a whole. I shall use this flag as a guide, a map to further explore the core's fluttering fabric of forces.

The flag's horizontal and vertical stripe of the cross of St. George shall, for the purposes of this text, represent the vertical rectus abdominis and horizontal

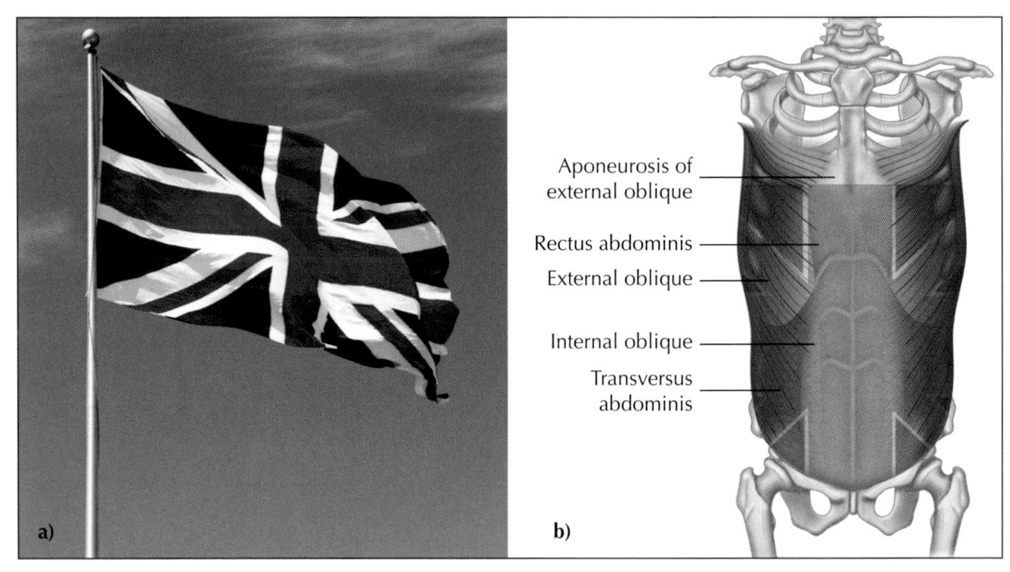

Figure 3.2. (a) The Union Jack flag; (b) the four bilateral muscles of the abdomen resemble a Union Jack, with the obliques forming an "X," and the rectus abdominis and TrA forming a cross.

TrA. The oblique cross of the red or white saltire represents the obliques.

The most satisfying part is that this anatomical pattern of muscles is repeated on the back of the body. The obliques "X" is replaced by the "X" of the upper and lower fibers of quadratus lumborum, the vertical rectus abdominis is replaced by the vertical erector spinae, and the TrA continues its loop to connect front to back. While the basic orientation or pattern of the anterior and posterior muscles is pleasingly similar, there are key differences in function and specific location.

The function of muscles is formed by evolutionary necessity. The anatomical map shows a crisscrossing pattern that exists to deal with the diverse forces to which our body is subjected.

While not a complete muscular picture, the pattern of the Union Jack is useful in understanding the need to exercise in multiple directions. Exercises that only focus on linear movement, for example, would increase the dominance to one part of the pattern. Such singular plane action would likely cause discord and imbalance, creating a dominance toward the more vertically orientated muscles represented by the vertical part of the cross of St. George. For some, the English are already too dominant a nation, just ask anyone who desires Scottish independence!

A more comprehensive understanding of the role of these muscles allows a more complete picture that can be applied to training as we develop a more balanced, strong, mobile, and adaptable multidirectional, multidimensional system (figure 3.3).

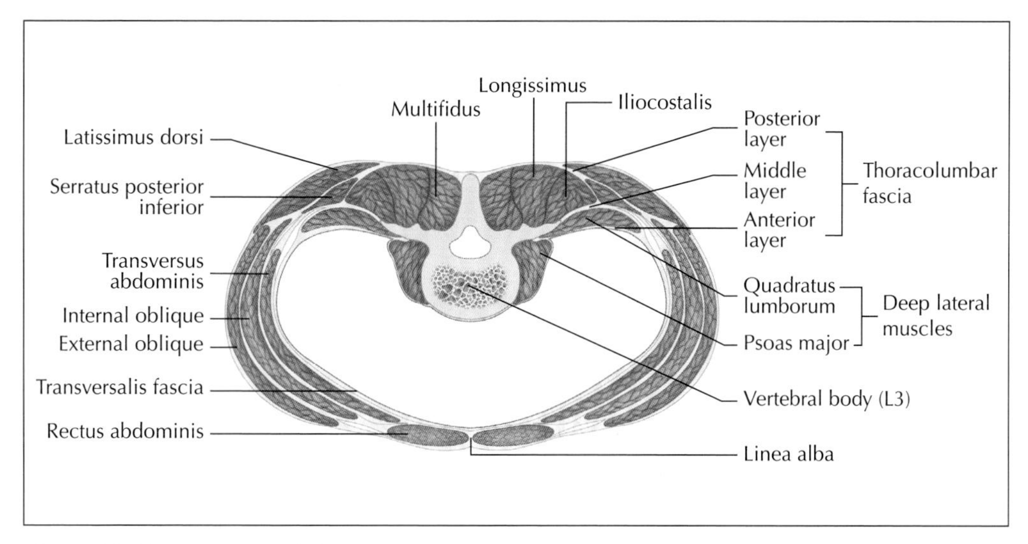

Figure 3.3. Cross section of the torso.

Separate books could be—and in some cases have been—written about each individual muscle associated with our core. Here, we shall briefly discuss some of these key muscles. I refer the reader to any number of comprehensive anatomy texts for a more complete and detailed explanation. The aim here is to find a practical use for some of the key principles described in anatomy texts.

The problem with discussing each muscle in isolation is that it increases our sense of its importance. Most of the muscles we discuss here are thin and subtle; a point driven home to me whenever I teach palpation to manual therapists. It is not unusual for manual therapists to have an overblown sense of the importance of a particular muscle. This perception can be conveyed into their hands—the result being a tendency to push too hard and miss the subtle differences of fabric texture and direction. This contrasts with the practitioners who are able to listen with their hands and feel the delicate layers organized like the fine layers of a croissant. Each layer is exquisitely thin, the thinnest superficial structure of the external oblique (2.81–3.17 mm) and only slightly thicker than the internal oblique (4.02–5.15 mm) beneath which lies the papery and deepest TrA (2.31–2.57 mm) (values from Rahmani et al., 2018). This is a significantly different experience that requires a subtlety that contrasts palpating the meaty Cornish pasty of the quadriceps.

It is clear that all muscles influence one's stability, mobility, and strength. Ideally, we should discuss all the muscles all of the time, but this would be confusing and unpractical. Fragmented learning, however, should not be confused with training. Exercises should focus upon movements and not on individual muscles. A fact emphasized by leading researcher Stuart McGill who explains that "it would be counterproductive to focus activation on just one muscle . . . training a single muscle appears to lead to dysfunction or at least compromised function" (Vleeming et al., 2007). McGill goes further, explaining the need to train movements, suggesting a "super binding" effect when all abdominal muscles are activated simultaneously, such that the measured torso stiffness is larger than the sum of their individual stiffness (Vleeming et al., 2007).

So it is with trepidation that I shall describe specific muscles as a recognizable mechanism for considering the complexity of the whole. Thankfully, this theoretical and fragmented approach to learning anatomy will not be reflected by the style of training advocated.

The Muscles

Rectus Abdominis

For many, the muscle most associated with the core is the rectus abdominis. Simply known as the "abs," it is the six-pack of Hollywood dreams and "perfect" beach bodies (figure 3.4). While the extreme of "chiseled abs" is not for everyone, improving the functionality of the rectus abdominis is a more useful goal for everybody. It is hoped that by taking a little time to appreciate this muscle, we will allow better training outcomes and improve the scope for more specific, variable, and interesting exercises.

Figure 3.4. In Hollywood even the monsters have six-packs. Although really it's eight compartments divided by three bands of connective tissue. The lower compartments are deep to other muscles (obliques and TrA) and so difficult to observe.

While we may think of the rectus abdominis as one muscle, it is actually two (figure 3.3). The right and left rectus abdominis is divided and so also connected by a band of connective tissue, the linea alba. An elastic tissue, the linea alba can expand during pregnancy and—in most cases—return to its original size postpartum. Together, the two rectus abdominis muscles act as powerful flexors of the torso in the sagittal plane. Most commonly the rectus abdominis is exercised by flexing the spine in a conventional sit-up. Reversed, and the rectus abdominis complies with the request to "tuck that bottom under young lady" by old-fashioned dance teachers.

Historically the singular and restrictive tendency has been to train the rectus abdominis only in the sagittal plane. The realization that the rectus abdominis is involved in side bending starts to alter our training. Simplistically, a contraction of the right rectus abdominis would cause a lateral side bend to the right and the opposite to the left. While I believe there is no such thing as a bad exercise, some exercises are poor choices for some individuals, tasks, or goals. Sit-ups often tell a story of poor functional understanding of movement that is based on cadaver dissection rather than real-world movement. Our current trend of movement training is still largely based on the actions of muscles as described in traditional anatomy textbooks. These actions are a result of the relatively crude method of pulling on a muscle and seeing its response at a joint or joints.

Previously we saw Dr Tulp (see figure 1.2) pulling the flexor carpi radialis to discover its actions. Pulling, or shortening muscles, such as the rectus abdominis, of a cadaver is the conventional method of movement discovery. Using this method, it is logical that a muscle that goes from the front of the ribs (ribs 5–7 and xiphoid process) to the front of the pelvis (symphysis pubis, crest, and tubercle) must flex the spine, which the rectus abdominis does. The next logical conclusion is that to train such a muscle, one should flex the spine against a force. The result of this logical method is the sit-up, an exercise that mirrors the actions of the supine cadaver.

Sit-ups help educate the rectus abdominis that its job is to flex the spine. But is this always the case? A simple experiment may help you to question this conventional logic.

Stand or sit and flex forward, bending your back. Now gently push into your belly—you'll find that your rectus abdominis is soft and relaxed not rigid, tense, or at work. Now try to extend your spine into a backbend, the opposite of the muscle action. Feel that your rectus abdominis has become harder now as the rectus abdominis fires and contracts.

This simple experiment clearly shows us that while the action of the rectus abdominis is flexion, its functional role is the opposite. The function of the rectus abdominis is to decelerate extension, so our training needs to better reflect the more normal functions of muscles. The rectus abdominis needs to be trained to lengthen under tension to become better able to control common actions that require controlled extension.

A lack of extension in the torso will likely result is more work elsewhere. Reach up for the cookie jar on the top shelf and the shoulders, for example, would be required to go into more flexion to compensate for a lack of spinal extension. This increase in workload often results in provoking overuse injuries to the shoulder, especially if you like cookies.

For most people, exercises that ask the rectus abdominis to resist or decelerate extension have more functional everyday importance than training flexion of the spine. If you run, walk, throw, or reach up to get the cookie jar, then it is logical to include extension exercises into your training (for example figures 3.5 and 3.6). If you only train flexion, you may well be reducing the functional capacity of the rectus abdominis.

Figure 3.5. An elastic band works arm extension and consequent pre-tensioning and work to maintain torso position.

Figure 3.6. Throwing and catching a ball. Torso tensions, decelerates, and recoils with each throw and catch of the ball.

Erector Spinae

If rectus abdominis is the Hollywood superhero, then the erector spinae (figure 3.7) is its nemesis. This long, ropey, often-forgotten workhorse of our daily battle with gravity, the erector spinae can be found on the reverse of our human flag. These strong cables—labeled back extensors—are a little more complicated and, I think, more interesting. As we discussed, the typical function of the rectus abdominis is to prevent extension, however the action of the erector spinae is to extend the spine and its function is to prevent flexion, the reverse of its action.

Rooted to the pelvis, the three cables of the erector spinae (spinalis, longissimus, and iliocostalis) reach up the back to attach at multiple points on our flexible ribs. The erector spinae is constantly working to prevent gravity from pushing us down into flexion. Increase the weight by holding a sandbag or something heavy in your arms and you'll soon become aware of the increase workload of the erector spinae (see figure 1.12).

The erector spinae acts like the guy ropes on a tent preventing our flexible (spinal) tent pole from bending forward. To be effective, guy ropes require a solid anchor, which is why a tent peg is driven into the ground. In the body, the pelvis acts as the ground, our solid base that is far from solid.

The pelvis itself requires motion to enable us to move effectively. Each of the three cables attach to slightly different

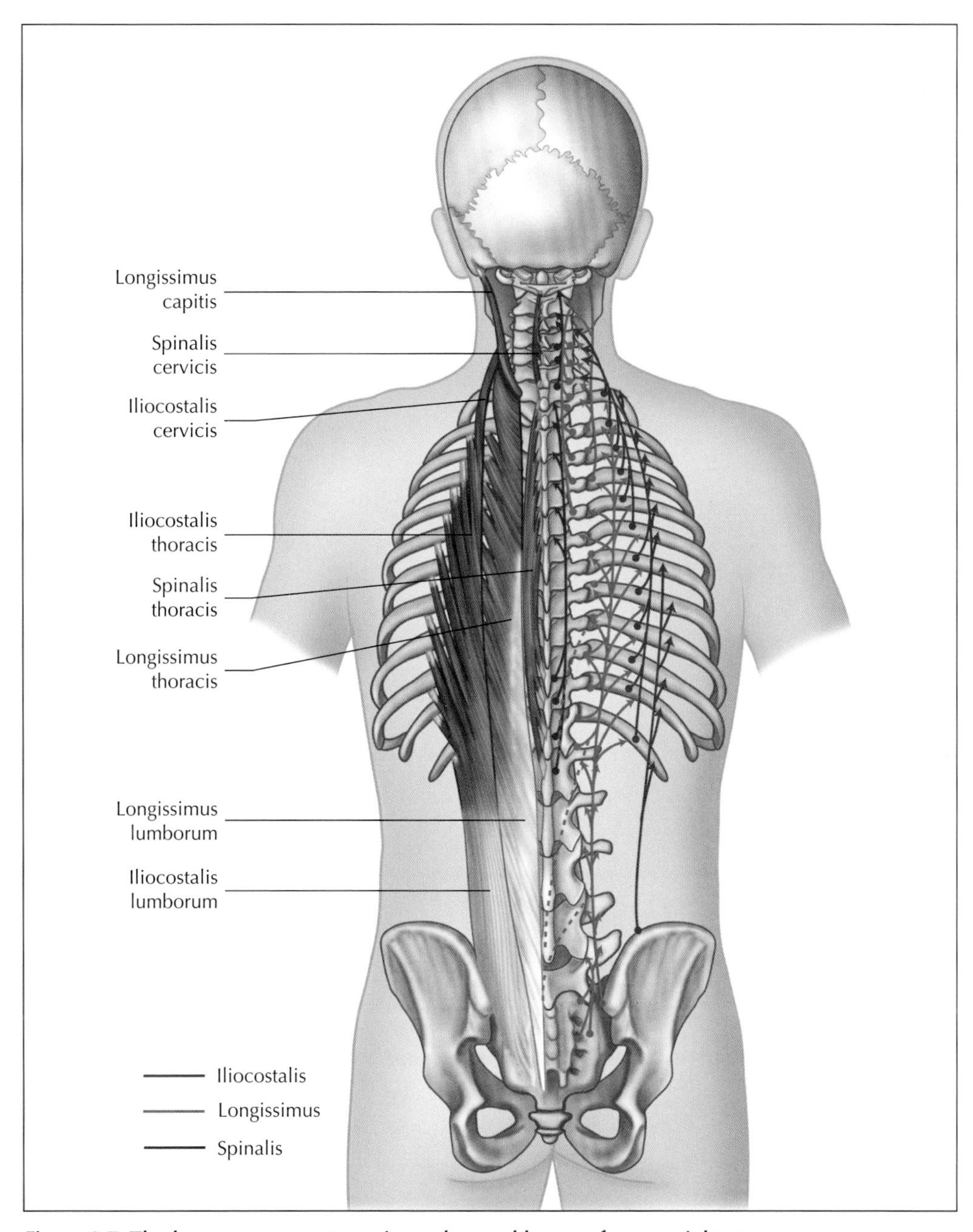

Figure 3.7. The long ropey erector spinae, the workhorse of our upright stance.

locations on the pelvis allowing each to perform different jobs relating to the rotations: torsions, nutations, and counter-nutations of the pelvis and sacrum (MacIntosh & Bogduk, 1991).

A chronically rotated or torsioned pelvis provides an unstable base for the erectors and their association with the ribs and spine. Twisted, the cables of the erector spinae can send issues spiraling up from the pelvis to manifest in dysfunction in the back, shoulders, neck, or head. It is always worth asking a professional to check that the pelvis starts and maintains its normal pattern of movement that is expected during any exercise.

Multifidus

If the spine itself has a core, then surely the multifidus is it (figure 3.8). Small and powerful, it deserves our attention, even if it is not part of our Union Jack of muscles.

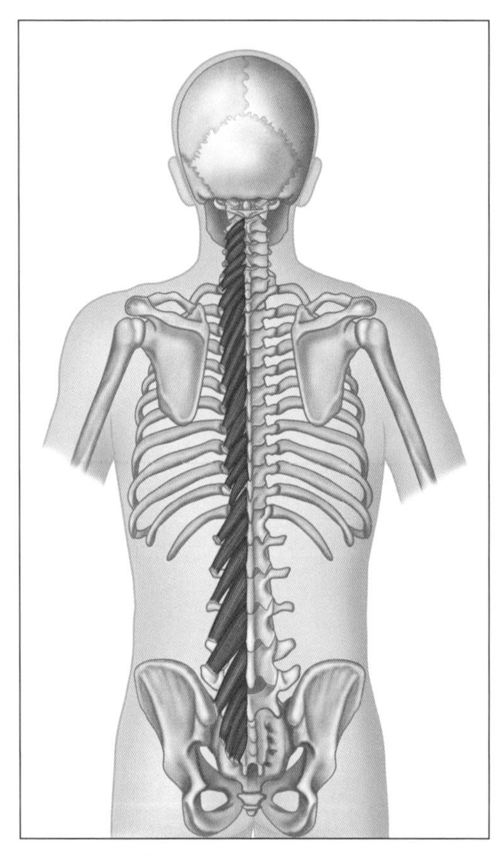

Figure 3.8. The mighty multifidus muscle is critical for a healthy back.

The multifidus is quickly establishing a reputation as one of the most critical muscles for a healthy back. The multifidus is the thickest muscle of the transversospinalis group that includes the semispinalis and rotatores. Among the smallest yet strongest muscles in our body, the multifidus is highly sensitive, with its sensitivity coming from the high number of muscle spindles. These sensory receptors detect any changes in the length of these short muscles. The short span of this muscle, two to four vertebral segments, gives the multifidus its strength. The rodlike arrangement of muscles fills the grooves between the spinous processes and transverse process of each vertebra.

As one of the stiffest fibers in the human body (Ward et al., 2009), the multifidus gives segmental stability to the spine. The reason for the strong link between the atrophying of the multifidus and lower back pain is currently unclear (Woodham et al., 2014). Like the chicken and egg debate, the uncertainty stems from not knowing if the lower back pain causes the multifidus to atrophy or the atrophying multifidus causes back pain. Whatever the reason, a healthy multifidus does seem to be associated with good back health.

If you have lower back pain, you are advised to find a skilled bodyworker to check your multifidus. Bodyworkers with good palpation skills can discover if the multifidus is strong and fires in sequence or is dysfunctional and atrophied.

With help, careful cueing, and a gradual graded development of awareness and activation, the multifidus's strength and functionality can be restored, often with wonderful results.

Finding exercises for this specific muscle group that are useful to the individual and correctly fit the individual's stage of development are critical for success. The exercises included in this text are more general and suited to those without back pain to act as preventive exercises. As the multifidus is almost continually active, most exercises could be included. While acknowledging that the exercises work on many other muscles, the examples—exercises such as figure 3.9—attempt to emphasize the multifidus's role in controlling rotation at each spinal segment.

Transversus Abdominis

Returning to our flag of muscles, let us focus on the TrA, represented by the horizontal line of the St. George's cross. Simply named by its transverse direction and position, this muscle is beloved by researchers and appears in multiple papers. The attempt has been to make this complex structure simple to understand—however, this seems to have added confusion rather than clarity. The TrA is not simply a horizontal band of tissue as its name suggests. The TrA is a multidirectional, complex arrangement of fibers that requires a training stimulus

Figure 3.9. Progressive exercises to emphasize the rotational role of the multifidus. Athletes can either be static and maintaining their position, or flexing and extending their arm.

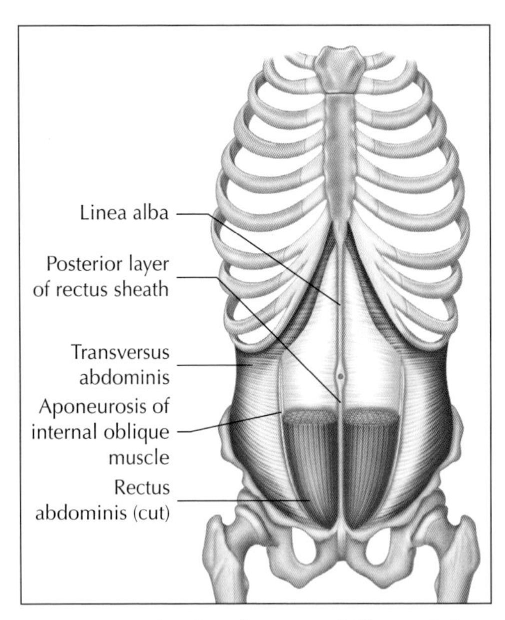

Figure 3.10. The thin, horizontal fibers of the TrA wrapping around the torso.

that reflects its variability in fiber direction and function (figure 3.10).

The thicker upper fibers give the TrA the directional derivation of its name. However, as we progress down toward the middle fibers, the orientation changes to be angled more medially. This orientation continues to be graded downward as we progress to the lowest and thinnest of the fibers. If you place your hand on your belly, fingers spread wide, and index finger horizontal, you should get a good impression of the gradual change in fiber direction. With its wide span, the borders of the TrA range from the inguinal ligament around the iliac crest to the thoracolumbar fascia (TLF) of the back, up to and around the inner surfaces of the seventh to twelfth ribs to the xiphoid process and linea alba at the front. Together, the TrA creates an enveloping bag-like structure around our middles. The change in orientation and thickness gives us a clue as to the role of the TrA and how best to train this complex muscle.

The inordinate quantities of TrA research often explores the TrA's role to act on the viscera and create stability to our watery organs. Another role of the TrA is its involvement in singing a soft lullaby or yelling to your teammate across the pitch. The TrA can control the different pressures needed in the abdominal cavity for these two styles of verbal communication. The TrA also takes a significant role in altering the IAP, which can enhance the hydraulic amplification effect to increase the efficiency of muscles.

While all fibers of the TrA contribute to the changes in IAP, the thickness of the upper fibers suggest a functional difference. The horizontal fibers of the TrA connect the rectus abdominis at the front with the spine at the back, creating a muscular ring. This muscular ring somewhat mirrors the pelvic ring below and the rings of ribs, spine, and sternum above (Lee, 2018). This pile of rings creates stability for the torso and pelvis while allowing mobility between each stacked ring, rather like Slinky-Dog (figure 3.11) from the film *Toy Story* (Lasseter, 1995). It is worth considering the implications on force transfer when one or more of these rings are not in a stacked position.

Of all the muscles, the TrA most closely resembles the corset by creating a

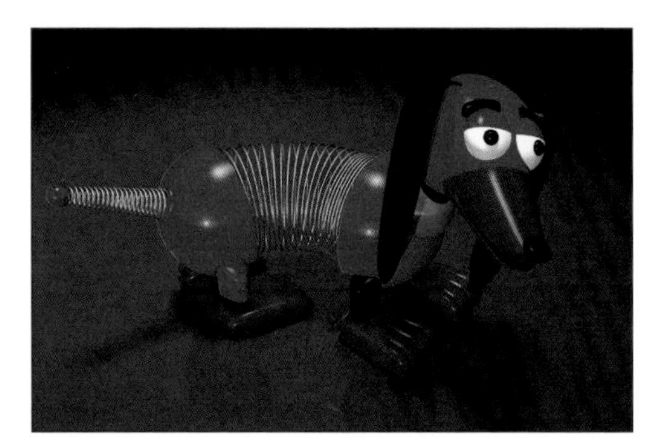

Figure 3.11. Slinky-Dog, from Toy Story *(Pixar).*

thin wrapper around the torso. As we have already discussed, the corset is a structure with significant psychosocial implications to our well-being. The TrA is a muscle with strong emotional ties. On hearing news of our beleaguered environment, the compassionate TrA will help you sob uncontrollably. If you feel unwell, our kind TrA will work overtime shifts as we retch, cough, defecate, and vomit, and it continues its work, subtly now, to help the light breath of restorative sleep.

Typically, researchers isolate muscles to study them, and the TrA is no exception. The consequence of this isolated approach is the prevalence of popular exercises that attempt to isolate the TrA. Taken together, the evidence shows that the deep muscles of the torso work synergistically, not separately. The TrA is ideally placed to disperse forces across its large surface area, which enhances its capacity for load distribution and to make subtle adjustments in stiffness. Instead of having a system that concentrates load, the human structure is able to decrease the workload on any single isolated structure. One of the fascinating characteristics of the TrA is that it fires or contracts prior to demand. This prepares the body for the potential increase in load by increasing the IAP and effects of the hydraulic amplification system. Added to this, the variety and complex arrangements of attachments, most notably to the rectus sheath and the TLF, allow the fibers of the TrA to deform and conform depending on the forces acting throughout the system. The fibers of the TrA are therefore capable of "simultaneous expansion and contraction along multiple axes" (Brown & McGill, 2008). These changes in the relative tension of the tissues allow the increase in the compression of the torso to maintain optimal joint position depending on the changing demands.

This complex synergistic relationship of all the fibers of the TrA is played out in a similar way throughout our body's inner relationships. The implication for training is that the more varied and multidimensional the forces, loads, and timing, the better the exercises are at targeting and educating the various roles of this complex, subtle, and perhaps kind, tissue.

Obliques

The next part of our anatomical flag takes us away from England and St. George to its neighbors in Scotland and Ireland. Both St. Patrick's saltire and the St. Andrews cross depict an "X" that is conveniently similar in orientation to be analogous with the internal obliques (IO) and external obliques (EO).

As its name suggests, the IO and EO lines are in a distinctly oblique orientation (figure 3.12). The aponeurosis (a thin sheet of tendon-like connective tissue) of the right EO is essentially continuous across the midline with the left IO. In the same way, the left EO is continuous with the right IO, resulting in the Celtic cross of unity. This "X" pattern led renowned anatomist Carla Stecco to describe the obliques as a digastric muscle (Stecco, 2014). A digastric muscle has two bellies but acts as one muscle, rather like a string of two sausages—the sausage meat being the muscle and the sausage skin the fascial sheath of connection and communication. In this way, four muscles (two IO and two EO) interlace to become two muscles and work as one functional unit.

With all these connections, it is worth asking why we make a distinction between the IO and EO. The answer lies in the relative depth of each muscle. The IO lies sandwiched internally to the EO but external to the deeper TrA. One may assume that whichever old Greek guy—(probably Greek, probably male) came up with the name—it was to explain the location and orientation to other dissectors. The name, obliques, successfully indicates its location, however, it fails to tell us of its function.

Coupled with the TrA, the EO and IO provide an elastic wall that helps to retain the visceral content within. The larger EO is especially well positioned to exert a compressive force on the torso necessary

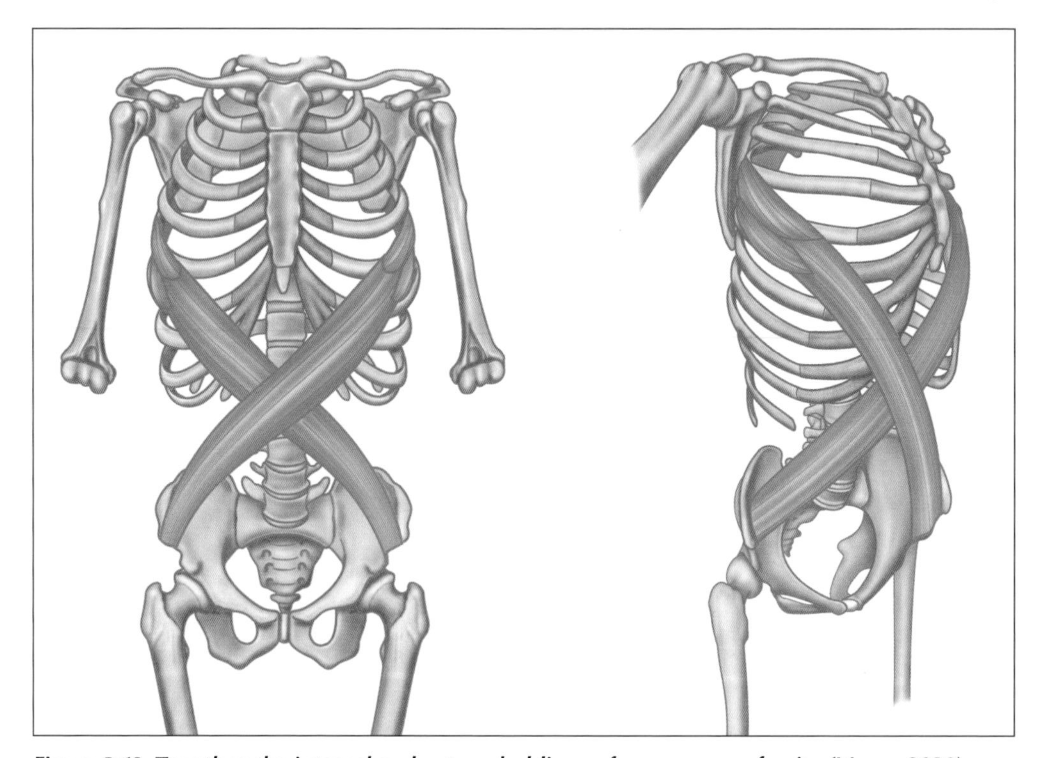

Figure 3.12. Together, the internal and external obliques form a cross of unity (Myers, 2020).

in expiration, defecation, childbirth, and vomiting.

Together, the right and left EO work to flatten the lower back into a common Tai chi position of readiness. The *Wuji* (or *Wu Chi*) stance translates to mean nothingness, a meditative "emptiness in any movement, thought or activity" (Liao, 2001). This version of "perfect alignment" is achieved by slightly flexing the torso and by a moderate or "soft" flexion of the knees, tucking in the pelvis and shoulders, letting the arms hang with hands slightly cupped, and the tip of the tongue pressed lightly to the roof of the mouth (figure 3.13).

One purpose of this stance is to bring focus to the lower *Dantian*, the energy center in Chinese medicine. While this may be an excellent position for martial arts, it may cause difficulties if held onto in everyday life. Over the years I have seen many devotees of Tai chi rigidly holding this "correct" position in everyday life. This is another example of poorly taught correctness driving rigid conformity through misplaced cognitive perception. This, like so many other correct postures, creates a solidity and certainty, obstructing the flowing capacity of a body to adjust. Ironically, Tai chi masters are said to use the watery flow of energy (*qi* or *chi*) by yielding as well as by action. Holding a solid correct posture seems paradoxical to the central idea of this "soft" martial art. The power of Tai chi in its fighting form is said to come from the idea of water. "Water is the softest thing yet can penetrate mountains and earth. This clearly shows the principle of softness overcoming hardness" (Tzu, 400 BCE).

Figure 3.13. Tai chi standing pose.

The *Yin* and *Yang* of Tai chi has, in my mind, a similarity to the balancing act of stability and mobility partly created by the obliques. The obliques can offer stability by decreasing the mobility of the ribs, pulling them down toward the pelvis. This stability strategy is an effective way to gain the solidity that is needed when dealing with large forces, lifting weights, or pushing cars that fail to start. If you hold this same level of tension and use this same strategy in the long term, it can become problematic and inefficient. This strategy compresses the lower larger lobes of the lungs and has the potential to cause long-term breathing issues. Such a strategy also has wider implications across the system, including the shoulders and arms, reducing the ability of the torso to extend, and so is likely to increase the workload elsewhere. The fibers of the EO interdigitate with either the pectoralis major, serratus anterior, or latissimus dorsi. Part of a continuous myofascial sling from abdomen to scapula, this sling continues right up to the neck and head, and down to the legs in one continuous piece of tissue (figure 3.14).

Lift your arms up as you exhale and note how far and how easily your arms traveled. Now repeat this experiment—lift your arms, but this time inhale, again note the distance and ease of arm elevation. For most people, breathing out as you lift your arms feels unnatural and

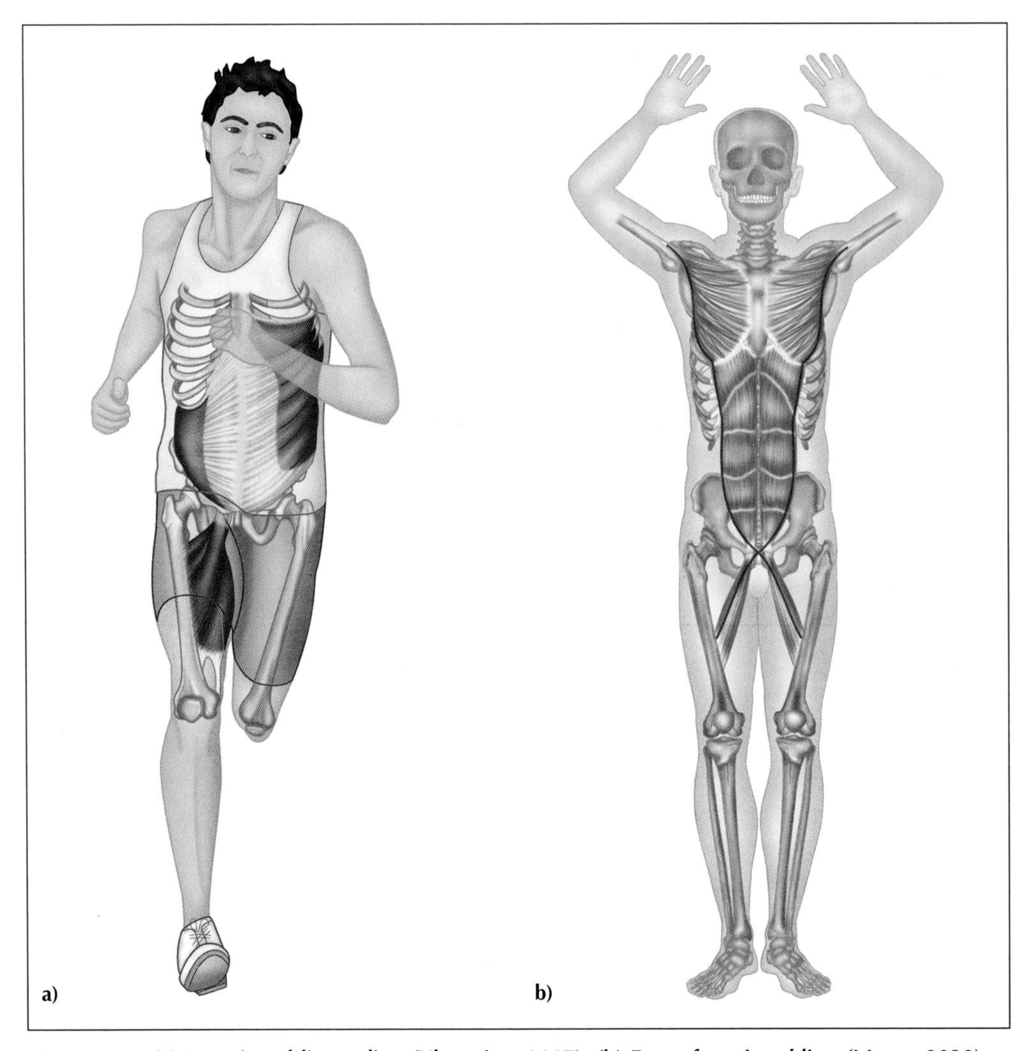

Figure 3.14. (a) Anterior oblique sling (Vleeming, 2007); (b) Front functional line (Myers, 2020).

takes a moment to think through. This exhalation strategy generally results in less arm motion and more strain. Lifting arms up and breathing in is generally easier, and most of us are able to go a little further and with more ease.

Forced exhalation is part of the job of the EO, which also causes flexion of the torso. When one needs to reach up, we also need to extend—not flex—our torso. The EO therefore needs to lengthen and control the extension, not shorten or contract into flexion. The reduction of movement capacity in one location (torso extension) will result in a greater demand elsewhere in the system.

Why are the obliques, oblique? Why do they have their characteristic angled orientation? What if they were straight? Part of the reason why the EO on one side is continuous with the IO on the other is that it determines the hollow curve of the waist. If the obliques were straight, then we would not have a waist. From a straight, waistless position, one would have to twist the upper rings of the ribs in one direction and the pelvic ring below in the other direction to create the waist as we know it. This describes the reverse of how our obliquely orientated tissue gets to the same situation in creating the waist—a "hyperboloid of revolution" (Gorman, 1981).

The oblique alignment plays a significant role in rotation. Their orientation allows the EO to turn the front of the ribcage toward the opposite side, and the IO turns the ribcage to the same side of its location. So, the left EO and right IO turn the ribcage to the right, and the right EO and left IO turn the ribcage to the left.

Turn to your left and as the ribcage rotates relative to the pelvis, one leg of the "X" shortens and one lengthens. The lengthening EO and IO allow two functions—(1) while lengthening, it will also decelerate and control the rotation of the torso, and (2) the elastic nature of the tissue will be stretched, become pretensioned and, like an elastic band, will recoil.

This windup-and-release mechanism acts like a watch spring, creating potential energy to release and recapture in a repetitive, rhythmical, efficient action of rotation and counterrotation, which can be felt in an easy efficient walking action. This elastic rebound-and-respond style of action is in contrast to an active contraction of muscles, shortening to create rotation. A more mechanical style of actively shortening muscles—when performed quickly and repetitively—is a calorie-expensive and inefficient way to move. Used in this way, muscles will weaken by overusing their capacity and, over time, will begin to produce less force. To increase the speed of movement with force, our bodies must use the elastic properties of the fascial bag. The muscles and momentum pull on this elastic tissue to rebound with an increased force and speed.

This understanding of the obliques and the fascial bag allows training to be an education in rhythmical motion versus

Figure 3.15. Use of a Swiss ball to emphasize the elastic qualities of our fascial system. Flex and extend the elbow and maintain the table-top position as rotational forces from the dumbbell change.

slow laborious contractions. Exercises that use props made of elastic materials, such as resistance bands, Swiss balls, etc., add to this education (e.g., figures 3.15 and 3.16).

Figure 3.16. An elastic band in one hand, playing catch in the other. A fun challenge to promote elastic rotation and coordination.

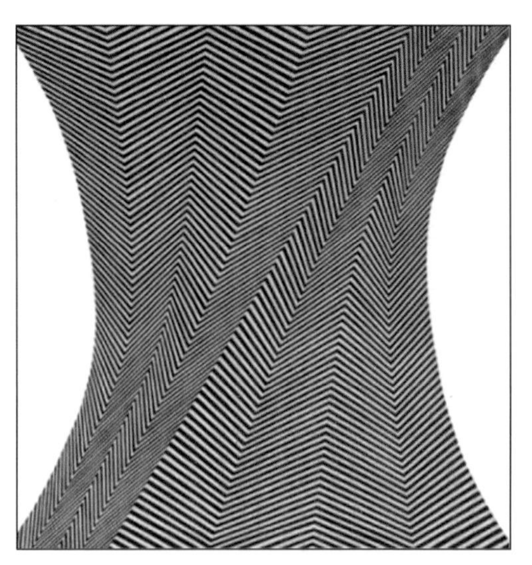

Figure 3.17. Stretch *by Bridget Riley seems to mirror our own torso and pelvic relationship and the "creation" of our waist.*
Stretch*, 1964*
Emulsion on wood
35 x 35 in
88.9 x 88.9 cm
(Cut to shape)

Later we shall explore how this is especially relevant for activities such as running and kayaking where finding a rhythm is key to gaining the advantages offered by the elastic qualities of this tissue.

I would love to think that Bridget Riley took her inspiration from the complex anatomy of the obliques when she created the image *Stretch* (figure 3.17). This image seems to encapsulate the waist and the relationship between the EO and IO. An angled web of relationships pulls the waist into the recognizable hyperbolic curve that corsetières attempted and Marilyn Monroe perfected (figure 3.18).

Figure 3.18. Marilyn Monroe.

Figure 3.19. Rory McIlroy showing off his exceptional rotational capacity.

Golf and Rotation

Throwing a ball for your dog, playing tennis, or a round of golf require the ability to rotate in a powerful and controlled manner. The tee shot in golf is a prime example of the need to find equilibrium between acceleration and deceleration and between mobility and stability. So let us briefly turn our attention to golf as it is such an excellent example of this movement principle.

A successful golf swing includes significant mobility, the rotational capacity of pro golfer Rory McIlroy is exceptional but not unusual (figure 3.19).

Increasing range leads to greater power and so a longer drive. Unsurprisingly, many golfers train to increase mobility and flexibility. Unfortunately, much of the mobility training—while wonderful for flexibility—is rarely specific to golf or another sport. In pursuit of mobility, the golfer finds themself in nonfunctional, long-hold stretches that do not look, or act, much like golf. While there is nothing wrong with performing a twisted down dog (*Parivrtta Adho Mukha Svanasana*, figure 3.20) all I am questioning is its functional relevance to golf.

It is well known that to increase mobility one must also increase stability. Note the use of three points of contact, creating much-needed stability for this amount of mobility. If you take away some of the stability, you will challenge the capacity for mobility. This understanding gives

Figure 3.20. Twisted down dog or Parivrtta Adho Mukha Svanasana.

Figure 3.21. A wonderful photo but one might question if this is the most functional, golf-like way to train. Photograph courtesy of Sandy Huffaker.

us a chance to explore the mobility and stability interplay and tailor it to the golf swing.

As an experiment into mobility and stability, swing a golf club and compare that with swinging the club while standing on one leg. This experiment into mobility could, with refinement, become part of your golf training. Did standing on one leg create a longer, more fluid, and graceful swing? Almost certainly not. Most likely you found that the lack of stability resulted in your whole body becoming stiff and rigid as it locked down searching for stability. Because of this lack of stability, it is likely that you didn't swing or hit the ball as far as when you were standing on two legs. In this position, the body struggles to decelerate the rotational movement.

During the drive phase, the golf club can accelerate to around 110 mph, causing massive rotational forces on the body. Thankfully, our body is expert in preventive medicine. To protect us from significant injury, our body works hard to decelerate this rotational force.

The interplay between stability and mobility means that a successful golf swing balances mobility with stability. The capacity to decelerate and protect is proportional to our capacity to accelerate the golf club.

Ultimately, training to improve mobility is also a question of stability. The body must allow an increase in mobility while in more unstable situations. The temptation here is to simply train on unstable surfaces, such as on a Swiss ball (figure 3.21).

However, similar to the debatable functional relevance of the down dog exercise for golfers, few golfers play golf on such uncertain ground. Pictures of golfers standing on large rubber balls is great for publicity shoots and injuries. To increase the range, we need a more stable base—like in the yoga pose—but one that might better reflect the golf swing. As legendary Canadian strength

Figure 3.22. Golf-specific lunge with step around to increase the range of the back swing with weighted golf club.

Figure 3.23. Rotational lunge exercise with weighted golf club to increase the acceleration and follow-through action for golf.

coach Charles Poliquin helps explain—you "can't fire a cannon from a canoe" (Poliquin, 1997)—we need a stable base for the mobility needed to exert power.

The work of David Tiberio at the Gray Institute has gone a long way to develop functional training for golf. One of his suggestions is to build mobility and stability by using a lunge-type action. The aim is to develop a golfer's ability to balance the key components of stability and mobility, acceleration, and deceleration within a golf-like action.

The exercises shown in figures 3.22 and 3.23 allow an increase in mobility and range by using a stepping lunge action while maintaining a stable and functional position. The result is an increase in speed and, in this way, we train the body to decelerate. This exercise increases the functional rotational capacity in the hips, pelvis, and torso. This improved ability to decelerate allows the subsequent increases in acceleration when returning to a normal swing (still recommended when not training). In time, this and other exercises will increase the range of mobility, the sense of stability, and, consequently, the power and speed of the golf club. While the exercises pictured use a weighted golf club, using a medicine ball, kettlebell, or elastics would add additional variation and challenges. For further exploration into this and many other functional golf-specific exercises, I highly recommend the Gray Institute.

Quadratus Lumborum

Quadratus lumborum (QL) sounds like something Harry Potter might utter and, like the teenage wizard, the "functions of the quadratus lumborum are a mystery" (Phillips et al., 2008).

Perhaps the reason that QL still mystifies is its multiple functions and fibers that fail to fit into one convenient category. This not-quite quadrate muscle is a woven blend of three layers of fibers into something more triangular-shaped. As the reflection of the rectus abdominis was the erector spinae on our Union Jack map, the reflection of the obliques is the QL. With exceptional skill, various dissectors have identified an almost cross-hatching of interrelationships that link pelvis, spine, and ribs, giving rise to its complex functions (figure 3.24).

The QL has anterior, middle, and posterior layers that run in distinct directions. Here we shall consider these three aspects of this one muscle and the implications for training.

The most anterior layer of QL is the most vertical of the three layers. These fibers connect the ilium with the ribcage, specifically the twelfth rib. Their orientation suggests involvement in flexion, extension, and side flexion. As with many of the core muscles, QL affects and is affected by breathing. Coupled with the iliacus below and above, the QL is a "direct extension of the diaphragm from twelfth rib to pelvis and can inhibit deep breathing" (Earls & Myers, 2017).

The middle layers of the QL connect the medial aspect of the twelfth rib with the lumbar spine. The angle of orientation suggests that these fibers have the capacity to move the lower ribcage toward the lumbars, side bending the torso. They are also well positioned to decelerate the torso, ensuring a controlled side bend when in the upright position of standing or sitting. Hold a shopping bag (or kettlebell, figure 3.25) in your left hand, and feel just above your pelvis as your right QL fires to prevent you toppling to the left.

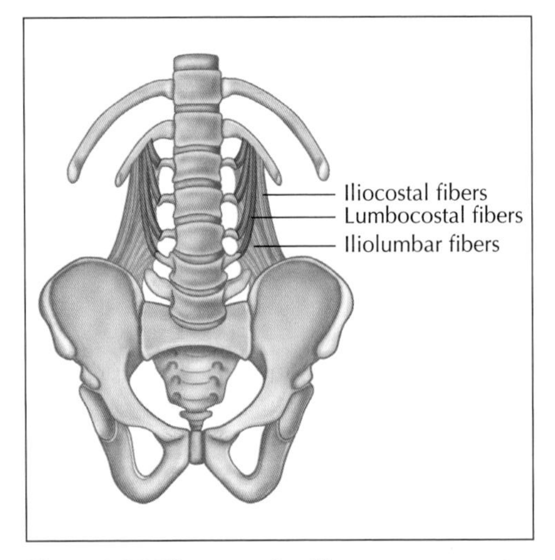

Figure 3.24. The complex fiber arrangement of the QL muscle.

Figure 3.25. Walking with one kettlebell for offset loading tasks.

As you place your shopping on the floor, taking care not to break anything, the right QL will carefully elongate to help decelerate the bag to the ground. Various training examples can be found in these pages that attempt to mimic and prepare us for numerous everyday, uneven loading tasks.

The pattern of the posterior QL fibers run from the ilium to the lumbar spine and "insert into the middle layer of thoracolumbar fascia" (Phillips et al., 2008). The slanted nature of these fibers indicates its role in actively moving the ilium toward the lumbars, resulting in tilting the pelvis to the opposite side to the contracting muscles.

Actively tilting the pelvis up on one side, while possible, is an uncommon everyday movement. The more common movement is the pelvis tilting in response to what is underneath it: the leg and foot. Step forward with your right leg and your pelvis will tilt to the left. That is assuming you took a long enough step and there are no restrictions to prevent this normal action. In this action, the function of these iliolumbar fibers (fibers that span the ilium to the lumbars) is to decelerate and allow—via lengthening—this necessary part of walking. An inability to perform this "hip drop" will decrease efficiency in gait, contribute to dysfunction, and cause more work and possibly pain or discomfort elsewhere in the system.

Lunges can be an excellent way to train these fibers to perform their decelerating task. A gym is a safe place to play with the direction (forward, backward, side to side, and everything in between), length (short, long, wide, and narrow), and rotation of the foot (front foot toed in or out, back foot toed in or out, both feet rotated, etc.). While the standard forward lunge is useful, performing a variety of lunges allows one to be better trained to deal with the variety of possibilities of taking a step. The QL is an excellent example of the need to train the omnidirectional, omnifunctional nature of human movement reflected in our anatomy.

Figure 3.26a & b shows our model standing and bending one knee to allow the pelvis to tilt. On the right side, the fibers of the right QL lengthen to control this action and allow the lumbar spine to remain relatively neutral. This is an example of optimal biomechanics.

However, something different is happening on the left side (figure 3.27a & b). Here the pelvis is tilting to the left, but the QL is not lengthening. The result is that it pulls the lumbar spine into a dysfunctional bend. The left side demonstrates suboptimal biomechanics. However, we should not infer pathology from this photograph: further investigations are needed.

The right and left QL operate in a synergistic relationship like that of the IO and EO. Together with the IO and EO, the QL helps to shift the torso laterally. If you've ever had the honor of holding a child for a long period

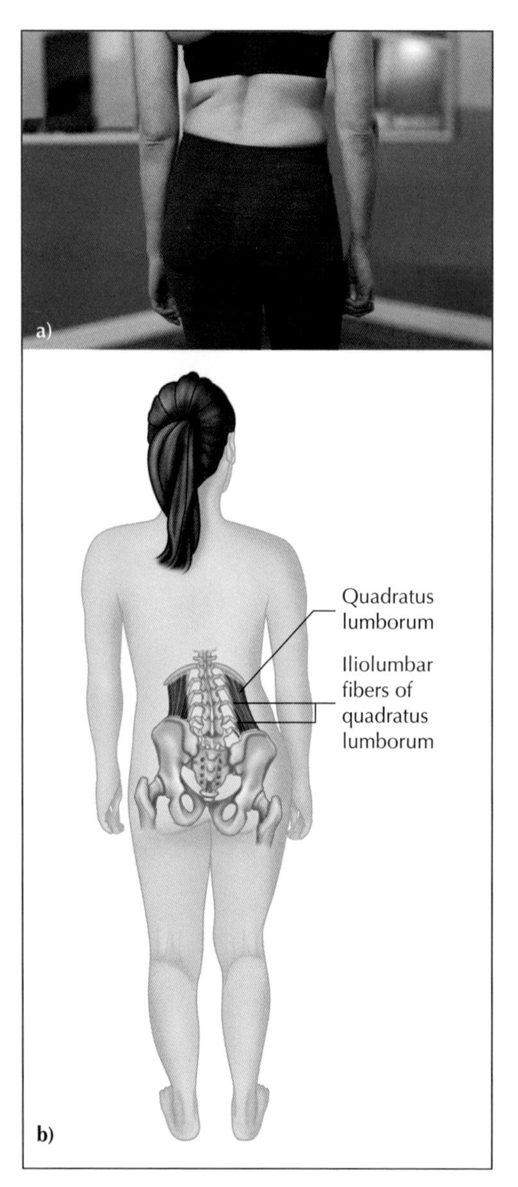

Figure 3.26 (a) & (b). Right tilt of the pelvis showing the lengthening of the right QL.

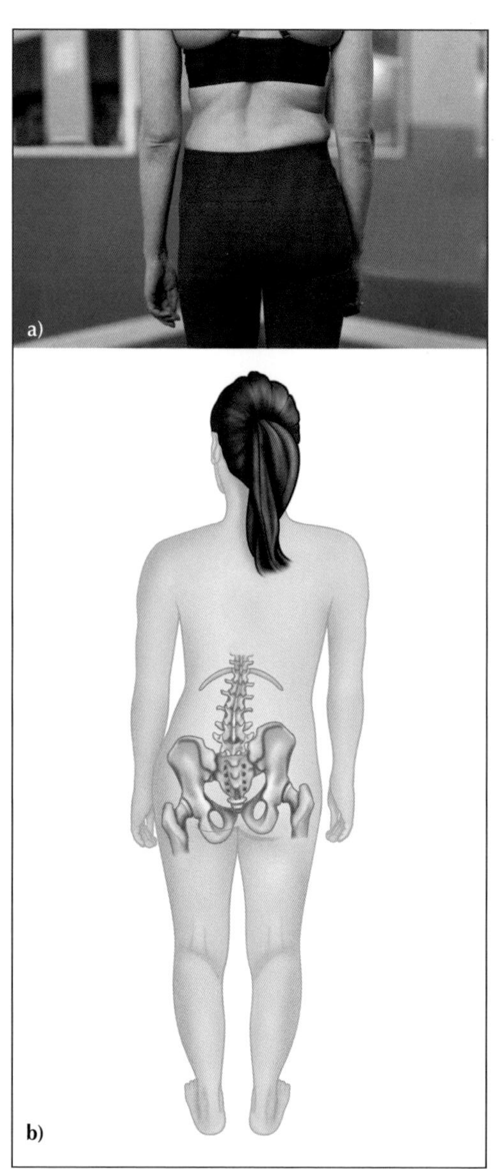

Figure 3.27 (a) & (b). Left tilt of the pelvis showing difficulty of the left QL to lengthen causing a bending in the lumbar spine. This is suboptimal and common.

of time, chances are you'll appreciate the work of the QL. Finding a way to hold a child and still make the dinner is a curious problem familiar to most parents. I, like many parents, solved this problem by placing my daughter on top of one side of my pelvis to free up the other hand for cooking. The QL's role in this culinary juggling act is to shift the ribcage to one side, allowing a larger

Figure 3.28. The Hippychick Hipseat allows the wearer to maintain a more neutral spine and pelvic alignment.

pelvic "seat" for the child. Shifting the ribcage to the right requires the upper middle fibers of the QL on the left and lower posterior fibers of the QL on the right to work together. Together they create a functional cross to allow the necessary shift of the ribcage.

It goes without saying that other muscles are involved, most notably the other "X" of the IO and EO. Many parents will testify to the discomfort caused while maintaining this functional, yet far from ideal, position. Thankfully for me, by the time my son appeared, I had discovered a simple and practical solution to this problem, the Hippychick Hipseat (figure 3.28). This simple and effective little hip seat allows the wearer's spine to maintain a straighter alignment to reduce the workload of the multitasking parent at feeding time.

Thoracolumbar Fascia

The background color of the Union Jack flag is blue and represents Scotland. The back of our core anatomy is fascial and

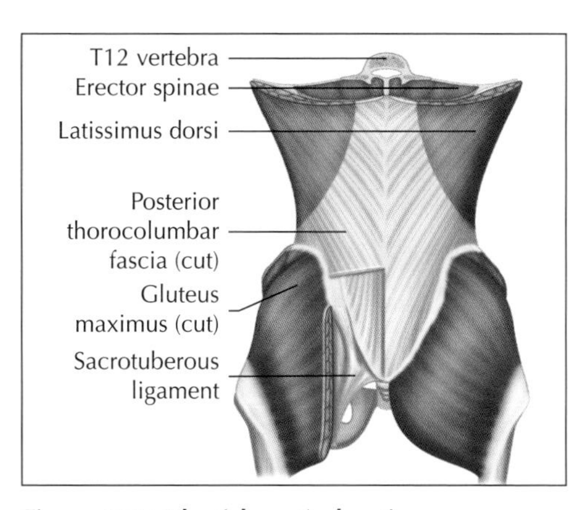

Figure 3.29. The (almost) chaotic, multidirectional fibers of the thoracolumbar fascia.

(seems to me) to represent chaos. Like chaos theory, the TLF is highly sensitive to slight changes in conditions. The most famous example of chaos theory is the butterfly effect (figure 3.29).

The mathematician and meteorologist Edward Lorenz originally used the less poetical seagull in a storm to explain how small changes—like the flap of a butterfly wing—can create a hurricane elsewhere in the world (Gleick, 1997). Tiny changes that have the potential for great effects. The location of the TLF, in the lower back—combined with the nature of its fabric—makes it a perfect receiver and responder. The TLF listens to the changes in forces from above and below, from the hurricanes of gross movements to the subtlety of fine butterfly-like actions.

For this reason, it is hard to think of a specific TLF exercise. However, multiple activities indirectly train the TLF.

A primary function of this whole abdominal region is to transfer loads from lower to upper or upper to lower body.

A long list of muscles attach into the TLF, including the TrA, IO and EO, erector spinae, all of which we have already discussed. The list also includes gluteus maximus and hamstrings (biceps femoris) of the legs and latissimus dorsi of the upper body. The variety of muscles indicates the ability of the TLF to transmit forces throughout the body. Transferring load diagonally from right arm to left leg and left leg to right arm is necessary to run, walk, throw, play racket sports, and various other contralateral actions.

With around eight times the body weight being transferred into the front foot on the final step, throwing the javelin is clearly one of the most dynamically forceful actions in all of sports (figure 3.30).

Figure 3.30. The explosive action of throwing a javelin.

Critical to success in throwing events, such as the javelin, is the capacity for our body to transfer forces. Poor biomechanics and alignment of structures can cause "energy leaks" where forces are dissipated out of the body and fail to be used efficiently. Throwing the javelin requires the energy that is built up from the run-up to be transferred effectively into the javelin. The moment when the energy from the run-up is converted is the blocking action of the final step. This blocking action, coupled with the straight knee, sends a whip-like transfer of forces from the floor into the hip, torso, arm, and javelin. Even a slightly bent knee will lessen the force being transferred and is an example of an energy leak. This straight-knee position is created and must be maintained by the muscles of the leg as the foot hits the ground.

In a similar way, when lifting a weight—for example, during a squat—forces are transferred from floor, through the legs and torso, and into a barbell. The TLF is ideally placed to ensure the torso position is maintained in a reasonably straight or neutral position to better transfer the forces. The muscles of the torso ensure the spine is not in a bent position, which would result in energy leaks, overwork, and inefficiency.

The significant role of the TLF was identified by Bojairami & Driscoll in the prestigious journal, *Spine* (2022). The research indicates the important role of the TLF in the dispersal and absorption of loads. This information has implications for the treatment and understanding of lower back pain and training. A growing body of literature now supports the notion that the fascial component of the TLF acts not only to support and transfer force but also acts as a messenger (Liem et al., 2017).

Fascia is a highly sensitive character in our story. The TLF contains a high proportion of complex receptors. These receptors include interstitial receptors, Golgi tendon organs, Pacini and Ruffini corpuscles. When stimulated, these receptors have the capacity to "decrease muscle tone (Golgi), provide proprioceptive and interoceptive feedback (Pacini and interstitial), inhibit overall sympathetic activity (Ruffini), and increase vasodilation and plasma extrusion (interstitial)" (Lee, 2010). To stimulate and awaken these messengers, light foam rolling and fascial manipulation should be employed. The emphasis here is on subtlety and lightness, not hard rollers or hands that obliterate, but a touch that stimulates the intelligence of the TLF.

As French surgeon Jean Claude Guimberteau observed when taking a detailed look at fascia, the "architecture of the sliding system, which is responsible for harmonious movement in the body, is completely irregular and apparently chaotic. And yet our observations confirm the existence of efficiency of this irregular, chaotic, and fractal system" (Guimberteau & Armstrong, 2015).

Guimberteau's pictures of fascia remind me of the chaos-inspired images created by Jackson Pollock (figure 3.31b).

It is clear that our training needs to mirror the chaotic nature of fascia and so our own random fabric. This seems particularly poignant when we consider training to help the TLF. Linear, single-plane training—for example, only repeating the same standard squat—will fail to engage all of this chaotic fascial matrix of tissues.

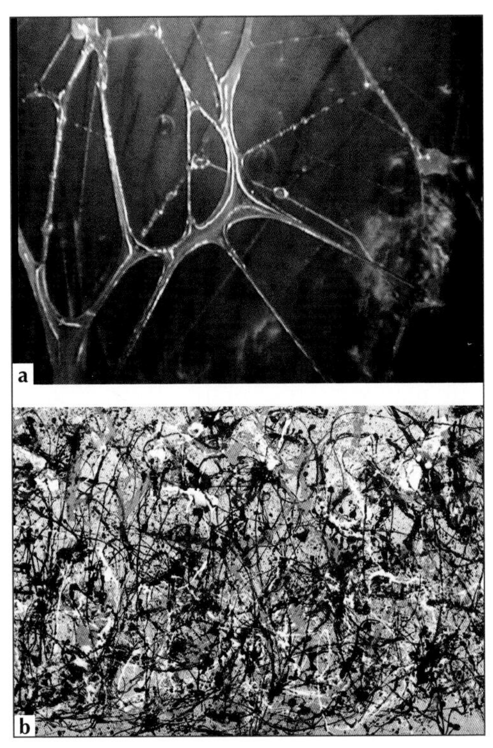

Figure 3.31. (a) The amazing photography of fascia by Guimberteau seems to mirror (b), the beautifully complex chaotic nature of the paintings of Jackson Pollock.

Exercise that is chaotic may suggest an "anything goes" approach, but this is not the case. The chaos of our tissue is "the hallmark of dynamic systems that do not conform to the rules of classical Newtonian physics . . . not the result of haphazard or random acts" (Guimberteau & Armstrong, 2015). It is this characteristic of "deterministic chaos" that allows training to have, and needs to have, a principle-based structure. Training that looks to load this chaotic fascial matrix must be in a variety of directions, using different weights, speeds, and using both bilateral and contralateral movements, for example. It is hoped that the imaginative athlete can develop other ideas to challenge, stimulate, train, and educate this intelligent and sensitive structure, but some examples can be found throughout these pages.

Psoas Major

Our anatomical tour of the Union Jack is almost complete—we have visited three of the four "home" countries. But Wales, the last on the list of UK countries fails to appear on the union flag. The official reason is that the flag was created in 1606 when the Principality of Wales was already united with England. However, I suggest the reason is far more poetical and far more Welsh. The British were simply not brave or romantic enough to prominently display a fairy-tale dragon on their flag. For this alone, Wales must certainly be the most mythological of all nations.

For me the muscle that this myth-loving nation must represent is the psoas (figure 3.32). Hidden from view, the psoas has gained almost folklore notoriety in certain circles. Unlike the dragon, the psoas does exist and—like the dragon—is of significant importance. Most agree that its actions include hip flexion, lumbar flexion, and lumbar support. Some debate remains to its other functions, allowing its mythological status to remain. This fan-shaped muscle is a muscle of connection as it "joins the upper body and the lower body, axial to the appendicular skeleton, the inside to the outsides and back to the front" (Earls & Myers, 2017).

With its connections from the lumbar spine to the top of the femur (lesser trochanter), one action is to create the ability to sit up from a lying position. Simple cause-and-effect thinking implies that sitting up more often will strengthen

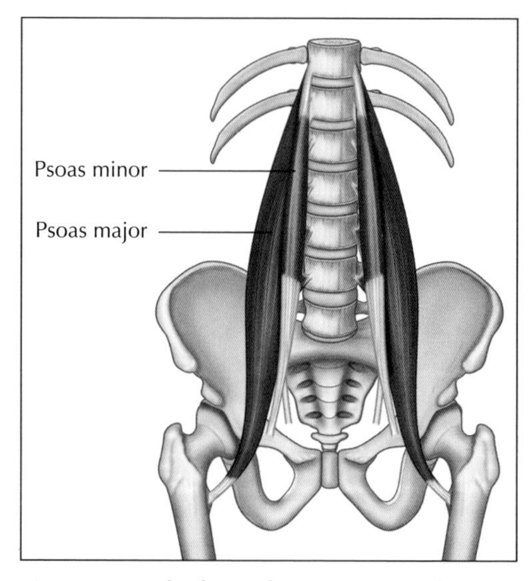

Figure 3.32. The legendary psoas muscle.

Figure 3.33. A conventional sit-up exercise.

and improve the functionality of this legendary muscle. However, it is now recognized that exercises, such as the sit-up (figure 3.33), promote maximal contraction of the psoas and "can exert a compressive load on the L5–S1 disc equal to 100 kg of weight" (Akuthota et al., 2008). The fact that sit-ups are regularly performed in gyms across the world without difficulty or pain is, I believe, testament to the resilience of the human body more than the brilliance of this exercise.

It is clear that shortening the psoas through exercise or lifestyle is likely to be detrimental to moving and functioning optimally. Often, we need to create length to counter our current prevalence for sitting, a habit that further shortens this, and other muscles. The psoas is "inherently susceptible to pathologically shortening (particularly in the older patients with a sedentary lifestyle or chronic immobilization conditions) and requires regular stretching to maintain normal tone" (Schuenke et al., 2020).

Figure 3.34. An inefficient way to stretch the psoas muscle.

Stretching the psoas is complicated by the multiple joints that it crosses and the multidirectional nature of these joints. It was once thought that a lunge or kneeling lunge would be able to stretch and create tension to the psoas. We now realize that the lunge alone is an ineffective method of tensioning the psoas as it fails to consider the impact of the spinal rotations and counter rotations of this action (figure 3.34).

The method shown in figure 3.35 allows the "winding-up" and tensioning of the psoas tissues. By understanding how this stretch works, it is hoped one might devise new or improve other such exercises.

The stretch starts in a conventional lunge position but with the toes of the back leg turned in, pointing to the heel of the front foot. This additional internal rotation of the back leg twists or winds up the psoas tissue from the lesser trochanter and upward. Flexing the knee of the forward leg will begin to increase the pull. However, if that was all one

Figure 3.35. A more efficient and effective way to lengthen the psoas muscle. With the back leg internally rotated, bend the front knee and reach forward and up with hands.

did, then the lumbars would rotate away from the front foot and "go for the ride" and the psoas would maintain its relative length. In a normal lunge, the distance between the lesser trochanter and the lumbars—transverse processes—remains fundamentally the same and there would be little or no stretch to the psoas.

The trick is to create a counterrotation for the lumbars. This is done by reaching forward with the hand opposite to the forward leg. So, if one lunges forward with the right leg, the left hand should reach forward. This does work but it is possible to increase the stability of the lumbar spine to further isolate the psoas. This is done by reaching the opposite hand up, adding a side bend component to the lumbar rotation. In a right-foot-forward lunge, the right hand reaches up causing a slight left bend, acting as a counter to the rotations. Explaining Fryette's laws of spinal mechanics is beyond the scope of this text, however, the practical value is evident in this explorative and expansive psoas stretch. For more on spinal mechanics, I suggest reading *Spinal Manipulation Made Simple* (Maitland, 2001).

CHAPTER 4

From Flag to Cylinder

A Tale of Two Domes

Let architects sing of aesthetics that bring Rich clients in hordes to their knees; Just give me a home, in a great circle dome Where stresses and strains are at ease. (Buckminster Fuller)

The analogy of a flag can only take us so far—we must add a third dimension to our core understanding. The torso is made up of two Union Jack flags (front and back) linked by the continuous concentric ring, the TrA, to enshrine our cylindrical structure that contains the viscera and functions to distribute and transfer forces. This cylindrical pressure cooker has a roof and a floor, which aids the multiple functions of this area. In this analogy, the roof of the pressure cooker is the diaphragm and the floor is the pelvic floor. We shall spend a few moments considering these two structures and their sibling-like relationship. The statement an "optimal pelvic floor function is also related to appropriate diaphragmatic activity" (Chaitow et al., 2013), can be reversed to describe the reciprocal relationship of diaphragm and pelvic floor. This interrelationship gives each structure its characteristics and functional capacity to enhance the multiple roles of the another.

The dome of the diaphragm and the inverted dome of the pelvic bowl are positioned such that they may communicate via pressure changes in the abdominal contents and differences in soft tissue tension. These two domes respond in a reciprocal relationship, the pelvic floor responding to the movement of the diaphragm and the diaphragm to the movement of the pelvic floor. The diaphragm, pelvic floor, and muscles of the abdominal wall affect the pressure in the abdominal cavity. The result of this collaboration is to reduce the stress on the spinal column as well as to stiffen the trunk wall as needed. This can be experienced during heavy lifting tasks, during which one may unconsciously hold one's inhaled breath. This inflated space acts to lighten the pressure load on the intervertebral discs by up to 50% in the upper lumbar spine and 30% in the lower lumbar spine (Schuenke et al., 2020).

During more restful periods of life, the alliance between diaphragm, pelvic floor, as well as other muscles of the abdominal wall act to gently massage and so improve the health of the visceral contents (Barral & Mercier, 1988). This assumes that the dome of the diaphragm and the dome of the pelvic floor are facing one another.

For me, the use of platforms like Zoom during COVID-enforced lockdowns highlighted the benefits of face-to-face communication. The diaphragm and pelvic floor also communicate best when facing one another. An example of a poor relationship between these structures is when the ribcage is tilted back and the pelvis tilted forward.

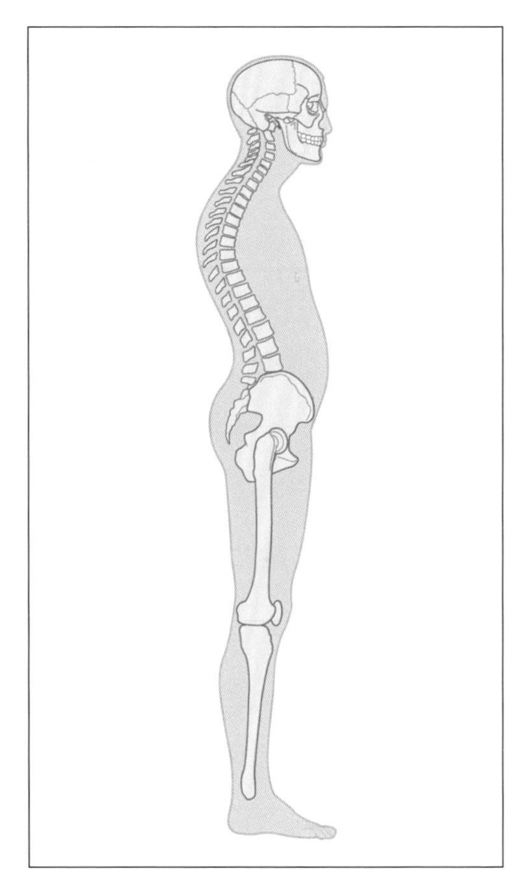

Figure 4.1. The common pattern of an anterior tilted pelvis with a posterior tilted ribcage.

In this position, the forces of the descending diaphragm on the inhale, and the ascending pelvic floor on the exhale, are directed not toward its reciprocal partner but out toward the thin belly wall (figure 4.1). This force, cascading into the abdominal wall some 22,000 times per day (average breaths per day) (*Canadian Lung Association*, 2022), overburdens muscles that are not arranged to manage forces from this direction.

As we have discussed, posture—and with it, body shape—is not necessarily a pathological condition; however, a particular shape can be undesirable.

It may be tempting to conclude from stories like the man with the beer belly, that a "correct" or "neutral" pelvis and ribcage would alleviate other conditions. As we have discussed, the idea that posture and pain are correlated is one that unfortunately persists. This persistence in our folklore—despite studies that found that 80% of males and 75% of females have anterior pelvic tilt and no pain

Case Study

A few years ago, a man came to my clinic complaining that however many abdominal exercises he did, he could not get rid of his beer belly. Having once been a regular drinker, he had totally stopped drinking 20 years ago and was now a regular runner and exerciser. He attributed his past drinking to his shape of an anterior (forward) tilted pelvis and a posterior (back) tilted ribcage. Together we worked to create length—via manual manipulation—to his chronically short-tight erector spinae muscles and hip flexors and rebalanced other associated tissues.

With some tweaks to his abdominal exercise program (including many in this book), I am happy to say that he achieved his goal—his beer belly shape was no more. The consequence was that he told all his friends and relations of our success, resulting in many of his beer-drinking friends visiting my clinic. Unfortunately, they were told to come back once they had achieved step one of the miracle cure—to stop drinking and eating excessively—and start exercising (sometimes it is that simple).

(Herrington, 2011)—means that we must therefore conclude that the once longed-for neutral postural position is not normal or an indicator of pathology or pain.

For most activities, maintaining a neutral position is not optimal or even possible. Movement requires coming out of neutral—it's like trying to drive your car in neutral—you'll not go very far.

Diaphragm

The roof of our abdominal canister is the diaphragm (figure 4.2). Usually described in the singular, the diaphragm is perhaps more correctly described in the plural. The two domes of the diaphragm can be correctly characterized as two separate muscles (Pickering & Jones, 2002).

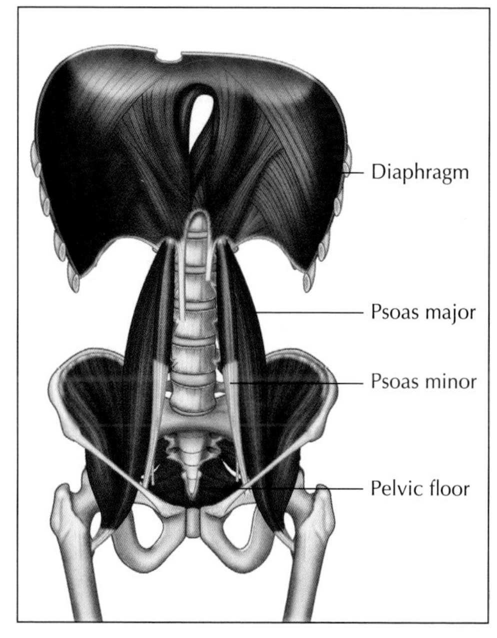

Figure 4.2. The diaphragm—the roof of the abdominal canister—shown here with the psoas connecting to the floor of the canister, the pelvic floor.

These two domes are in an ideal position to act as a pressure generator to affect IAP. The accessory muscle of respiration is of particular interest to us as many are located in the torso.

For example, the TrA and obliques work in harmony with the diaphragm to create ideal IAP. Ideal IAP coupled with the "corseting" of these muscles is part of the mechanism of spinal stability that helps protect the lumbar spine. "Deviation away from the optimal trunk lumbo-pelvic recruitment pattern, such as the oblique muscles firing first, then pressure, ventilation volumes, stability, and ultimately the work of breathing can be affected" (Chaitow et al., 2013).

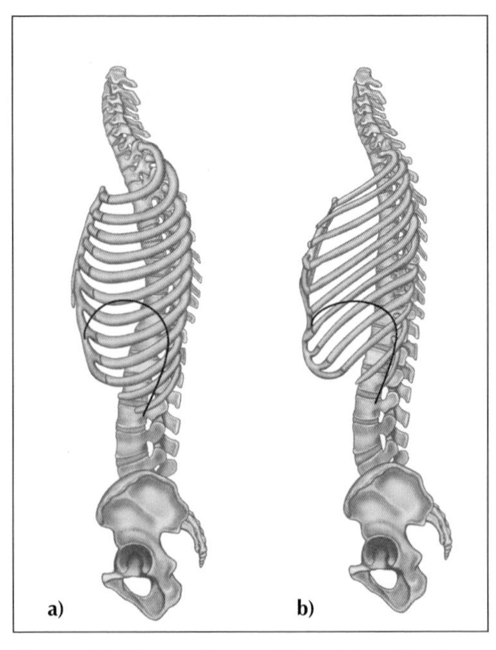

Figure 4.3. Zone of apposition: (a) optimal and (b), suboptimal positions.

Unfortunately—in the exercise world—this idea became misconstrued and led to the concept of bracing. Bracing—the conscious pre-contraction of muscles in preparation for load—is often associated with holding one's breath. It is now recognized that "this learned bracing will ultimately reduce stability: the TrA does not work in synchronization with the diaphragm: the zone of apposition [the area of attachment between the lower chest wall of the diaphragm and the ribcage] decreases, along with the ultimate force-generating ability of the diaphragm" (figure 4.3) (Chaitow et al., 2013).

In its place we all need to train the capacity to "change breathing pattern and rate, to meet the changing physiological demands" (Chaitow et al., 2013).

The capacity to be adaptable gives humans an "economical advantage" (Chaitow et al., 2013), which is perhaps easiest seen, and most researched, in the sporting context. A runner can "switch breath to stride ratio depending upon ventilatory demands" (Polemnia, 2007).

"Runners at low speed may take one breath to two strides (1:2 ratio) with higher speed switch to one breath every four strides (1:4 ratio)" (Polemnia, 2007). This is unlike a dog, whose breathing rate is coupled to the rate of their limb movement via muscles of breath. Muscles such as the serratus anterior are both muscles of breath and limb movement and so both movements are inextricably coupled (figure 4.4).

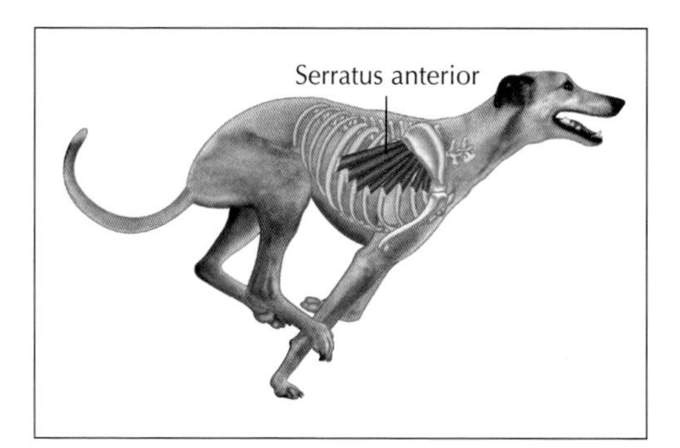

Figure 4.4. A running dog highlighting the dual role of serratus ventralis (serratus anterior) for breath and movement.

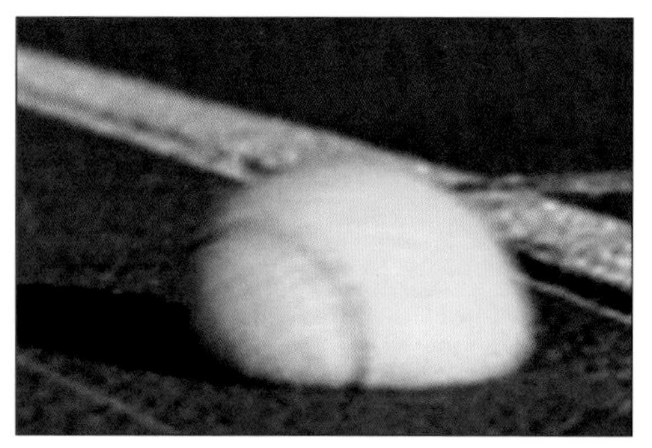

Figure 4.5. Bouncing ball.

This coordination of breathing to the rhythm of motion is not a skill confined to running: rowers exhibit similar actions. A rower may exhale once during a stroke, when the oar is in the water, and inhale once during the recovery phases, when the oar is out the water—a 1:1 ratio. As the stroke rate increases, this ratio changes to 1:2 with one complete breathing cycle during the recovery phase and one complete cycle during the stroke phase of rowing (Steinacker et al., 1993). It seems likely that other activities also use our ability to adapt both our muscular and breathing response to generate IAP, which enhances force transfer and efficient movement.

A constant theme throughout this book is variability and adaptability in training. The need for this is further highlighted as we consider the relationship of the breath to the abdominal muscles in more depth.

Poor training of the abdominal muscles overemphasizes the concentric shortening of the muscles. The result of such training is often a solid muscle with a limited capacity to adapt, lengthen, or decelerate actions. Imagine a ball made of concrete: it would fail to function as a ball because it would not bounce. Conversely, a rubber ball can use its elastic capacity. As this functional ball hits the ground, it morphs in shape, lengthening and increasing the tension on its elastic tissue, thereby increasing the potential energy of the ball to rebound (figure 4.5).

A similar length-tension relationship is necessary for a functional abdominal canister to breath (not bounce). Taking an easy relaxed "belly breath" allows the diaphragm to descend and the abdominal wall to enlarge as the pressure changes. Such a breathing style is one of rest and recovery as advocated by various schools of meditation. Breathing down into one's belly has significant psychological, as well as biological, benefits. However, if the abdominal muscles are in a short, solid state, the diaphragm may be restricted by the inability of the viscera to be squashed and hydrated. The result is the reduction of the IAP generated by the diaphragm. This pressure can determine the optimization of the length-tension relationship of these muscles. Shortened muscles create less force; the muscle's length-tension relationship is altered and the work of breathing increases. This can cause an uneconomical method of breathing that may cause chronic overbreathing. Such breathing pattern disorders have been linked to chronic fatigue (Nixon, 1994), neck, back, and pelvic pain (Smith et al., 2006), fibromyalgia (Naschitz et al., 2006), anxiety, and depression (Han et al., 1996).

Most of our breathing is an unconscious act—however, the effect of the diaphragmatic action is not some alien mechanical action divorced from the rest of our system. Breathing affects us both biomechanically as well as physiologically. Breathing rates and depths make significant changes in our blood chemistry, directly affecting our health. As James Nestor explains in his book *Breath*, "there is nothing more essential to our health and well-being than breathing" (Nestor, 2021) and Lewit states, "If breathing isn't normalized—no other movement pattern can be" (Hawkins, 2016).

While true that breathing and health are coupled, it is perhaps the bi-directional relationship of breathing and psychology that is most profound. The anxiety you feel before a job interview, the moment in the interview when you "choke," the depression as you fail to get another job—and any number of similar examples—have been shown to have close association with breathing pattern disorders. Breathing pattern disorders have long been recognized and researched. Older still is the skill of breath training. Numerous forms of breath training are central to various forms of martial arts, yoga, and meditation, which use our psychology to affect our breathing and our breathing to affect our psychology.

A human being is only breath and shadow. (Sophocles)

Like many, I consider breath training to be of great physical and psychological importance. Correct breathing—like correct posture—is a difficult concept, partly because "ideal function models or norms for either breathing or human postural patterns are not universally defined" (Chaitow et al., 2013). It seems to me that adherence to a "perfect" breath is likely to encourage a rigid conformity to other ideals that is in direct contrast to the ethereal flowing nature of our breath.

Various "schools" of breathing and experts of breathing exist yet there is "no consensus regarding whether a given individual's breathing pattern or posture can be assessed as ideal or incorrect" (Chaitow et al., 2013). Breath training—like all training—needs to be individually tailored. Optimizing an individual's breathing pattern can enhance—in all dimensions—the capacity of their breath for their lifestyle. Such training can be

Figure 4.6. A demonstration of a hypopressive breathing exercise. Simpler beginner exercises are also available from a qualified practitioner.

easily accessed in various books on the subject, such as *The Oxygen Advantage* (McKeown, 2015), or simply spending time learning the skill of breathing exercises. With the help of a trained teacher, hypopressive exercises are an excellent method for the prevention of, or recovery from, urinary incontinence, pelvic organ prolapse, and other related disorders (figure 4.6).

Pelvic Floor

At the floor of our abdominal canister, we find the pelvic floor (figure 4.7). Here again we find a muscle that is poorly named—in terms of its orientation, the pelvic floor is not truly a floor. Altering our perception of the various roles of this group of twelve striated muscles that we call the pelvic floor may alter our relationship to these often-ignored muscles.

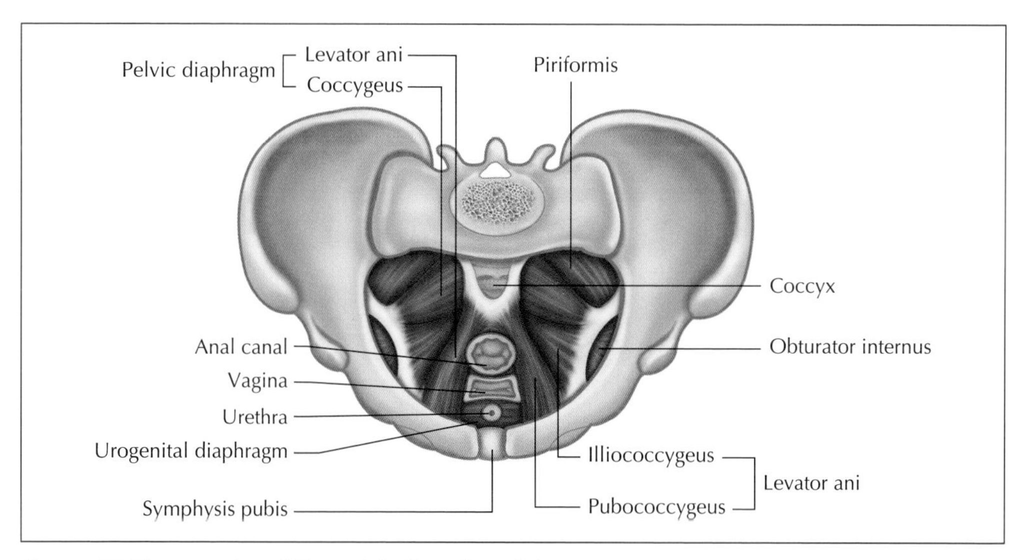

Figure 4.7. The muscles of the pelvic floor (female).

Various authors have suggested that the pelvic floor should be renamed the *pelvic wall* or *pelvic diaphragm*. While a name change is unlikely, pelvic diaphragm does remind us of the inextricable connection between the pelvic floor and diaphragm. But before we ditch the term pelvic floor, let us acknowledge the merits of this term. A floor gives us a sense of groundedness and connection that is consistent with its role as a root of support. The pelvic floor acts as a platform to support the three "poles": the spine, and two femurs. It is a platform that allows an area of connection, leverage, and efficiency of movement.

Tightness and/or weakness of the hip flexors is often attributed to prolonged periods of sitting. If sitting is the true cause of your hip dysfunction, simple stretches of the hip flexors can go some way to alleviate the associated symptoms. However, if a lack of support is the issue, then the hip flexors are working with an unstable (pelvic floor) platform causing inefficiency, overwork, and strain. This is a significantly different cause, but one often perceived as tightness or weakness that stretching will not alleviate.

The first time I tried to stand up on a surfboard was off the chilly coast of Northumberland in the UK. The instability of the surfboard caused my body to work hard as I spent hours desperately wobbling in a vain attempt to prevent another cold bath. The result of all this effort was tired, stiff, sore muscles and dreams of a holiday by warmer waters. Standing on an unstable platform—like a surfboard—results in overwork that eventually leads to inefficiency and poor function. If hip tightness is due to such pelvic instability, then it will not be until the pelvic floor (and other associated muscles) can function optimally that the associated tight or weak hips can start to improve.

For many, the pelvic floor is a vague area of limited understanding often by a naive shyness of "down there." Rarely discussed openly, the pelvic floor is often missed out in physical exercise. Perhaps this is of little wonder with terms such as urogenital diaphragm, urethral sphincter, and the anal sphincter, which are all associated with the pelvic floor. The importance of the pelvic floor means many of us will—as we age—need better understanding of "down there." While discussion is useful, exercise is perhaps a more powerful tool to bring the pelvic floor out of the shadows of conservative thinking and into a functional enlightenment.

Urinary incontinence or simply leakage is more common than most of us realize. The prevalence of incontinence is worrying, one study found that 95.3% of women with lower back pain also exhibited pelvic floor dysfunction (Dufour et al., 2018). It is worth noting that research often fails to determine if the lower back pain caused the pelvic floor dysfunction or if the pelvic floor dysfunction caused the lower back pain—either is possible, as are other causal factors. These worrying statistics are made more remarkable when we realize that estimates are based on under-reporting arising from social embarrassment.

In my clinical experience, it is not uncommon to find that a twist elsewhere in the body transfers an imbalance down through the body to cause a twist in the pelvis. This twist in the pelvic bowl can cause a muscular imbalance in the pelvic floor muscles that causes pelvic dysfunction and a variety of possible symptoms, including incontinence. Only last week I discovered that it was a client's old neck injury—from an almost-forgotten car accident—that was the root cause of his pelvic dysfunction and incontinence. For this client, no amount of pelvic exercises would ever improve his symptoms. It was not until the cause—his neck dysfunction—was identified that significant change could begin. Identifying the root cause of a complaint is a complex process and takes clinicians years of training to begin to comprehend. However, sometimes it's a much simpler process.

It has long been recognized that there is a link between urinary incontinence and pregnancy. The change in a woman's body and the increased weight of a wriggling unborn child on the bladder is often the cause for around 48% of women to experience leakage in the last trimester (Wesnes et al., 2007). For many women, the problem soon resolves postpartum. For an unfortunate 45% of women, this problem persists up to 7 years after giving birth (Lee et al., 2008). The fact that leakage is so common will, I hope, inspire pelvic floor exercises to become a normal component of pre- and post-natal education.

One might imagine that leakage is confined to people who are sedentary, pregnant, or suffer from back pain—unfortunately, this is not the case. There is growing evidence that even the fittest individuals in society experience leakage. Various athletic endeavors increase the demand of the IAP to stabilize and transfer forces through the abdominal region. This creates a greater need for the pelvic floor muscles to be elastic, flexible, and strong to deal with this increase in pressure. As we shall discuss, overactivation often creates rigidity and weakness. In elite-level sports, these pressure demands can overload the capacity of the system with unfortunate results. Studies focusing on female teenagers suggest that around 67% of gymnasts, 66% of basketball players, and 50% of tennis players regularly experience leakage (Nygaard et al., 1994). Remembering the significant positive physical and psychological effects of exercise, it is perhaps most worrying to note that around 20% of women reduce their activity levels due to leakage and 10% stop altogether (Salvatore et al., 2009). While most studies seem to focus on women, one would suspect a similar story is mirrored in the male population.

In light of such startling statistics, one might assume that there would be an extensive variety of pelvic floor-related exercises. Exercises that could be specifically tailored for lower back pain sufferers, women pre- and post-pregnancy, and the various demands of sports. The reality is that many people are limited to one set of exercises: Kegels.

Developing from the work of Margaret Morris (1936), who described tensing and relaxing the pelvic floor muscles, Arthur Kegels (1948) popularized exercises now commonly referred to as Kegels. As part of a daily routine of exercises, Kegels include initially identifying the pelvic floor muscles by stopping urinating mid-stream (please note that this should be an infrequent test, if repeated it can result in bladder infection). Once identified, the exercises then progress to imagining one is lifting a marble by tightening one's pelvic floor, holding for three seconds and then, importantly, relaxing. Kegels noted that his exercises should be performed without holding one's breath or contracting other muscles, such as those of the abdomen, thighs, or glutes.

The simplicity of these exercises led the National Institute for Health and Care Excellence (NICE) to recommend a trial of supervised pelvic floor exercises as a first-line treatment for urinary incontinence (NICE, 2019). While Kegels are no wonder drug of incontinence, they do represent a discrete exercise that is widely publicized in both women's and, more recently, in men's health magazines (Buttaccio, 2018). The popularity of these exercises, perhaps due to their ease to perform coupled with their catchy name, may at least initiate a discussion concerning the health of the pelvic floor. It is now clear that the effectiveness of Kegels exercises depends upon the cause of the underlying physiological condition.

While Kegels remain useful for some, they are not necessarily the answer and may make a problem worse. Unfortunately, this is particularly true for those who are most in need of help. People with pelvic floor dysfunction have been found to be significantly less likely to learn to do Kegels correctly on their own (Bump et al., 1991). These findings show that part of this problem is the potential issue associated with an overactivation of the pelvic floor muscles. Kegels are designed to activate muscles that are thought to be weak and dysfunctional. However, this is not always the case—for many, the pelvic floor is overactive, not weak. Activating an already overactive muscle will reduce its stamina, strength, and consequently reduce the ability to coordinate and time when to contract and when to relax.

As we have previously discussed, the result of some strength training is a solid, inflexible, unadaptable, functionally poor structure. The pelvic floor is not unique—like all muscles, the pelvic floor needs to be mobile, flexible, and adaptable to function correctly. The muscles of the pelvic floor must function like a trampoline to act and react to

Figure 4.8. Sitting on a foam roller to relax and "let go" of pelvic floor tension. Taking a deeper breath will increase awareness of the reciprocal relationship between breath and pelvic floor.

forces from above (diaphragm) and other directions. If the weakness is caused by over activation, then strategies to relax and education on how to "let go" is more applicable than strengthening exercises (figure 4.8).

It is important for individuals who exhibit symptoms of pelvic dysfunction to find a pelvic floor health specialist. Evidence is conclusive that the effectiveness of pelvic floor exercises increases when one is taught and supervised by a professional (Bump et al., 1991). Such professionals can consider the effects of diet and lifestyle, as well as psychological and physical stress on the pelvis. Pelvic health professionals who take a multidisciplined global approach are invaluable in the recovery from and prevention of pelvic dysfunction.

It is clear that pelvic floor exercises should be part of a complete package of exercises for the whole pelvic girdle, the whole core, and, ultimately, the whole body. Exercises to activate this area in a variety of situations and demands should be promoted. Such variety of exercises allows the training of the deep lateral rotators, piriformis, as well as the gluteal and other abdominal muscles, all of which influence the balance and functionality of the pelvic floor directly or indirectly.

One might speculate that, with the gradual movement away from specific muscle training toward a more global approach, we have come full circle to the pelvic floor exercises of 6,000 years ago. These ancient exercises attempted to awaken the spirit of this area rather than work any specific muscle. The routine of "deer exercises" in Chinese Taoism or the Ancient Indian *Ashwini Mudra* or horse gesture practiced by Yogis suggest a common appreciation for the importance of the pelvic floor. It is thought that the horse and deer were chosen as they represented health, stamina, longevity, spiritual development, and sexual well-being associated with a functional pelvic floor (Cohen, 1999).

CHAPTER 5

Putting Theory into Practice

Principles-Based Training

> *As to methods there may be a million and then some, but principles are few. The man who grasps principles can successfully select his own methods. The man who tries methods, ignoring principles, is sure to have trouble. (Emerson)*

Why does this book exist when the list of exercises available online seems to grow exponentially? It is only the internet that can keep up with the unceasing evolution of exercises. It would be impossible and impractical to include all exercises in this or any one book. The reason that this book can exist is that it is a book of principles not a list of exercises that will soon be out of date. I sometimes question the motivation behind some exercises I see online. Exercises have become ever more radical ways to entertain and act as "click bait" rather than as examples of well-considered movements.

The internet is a rapidly changing entity, while books are finite, carefully created, and static. At first glance, books don't seem to fit the fast-paced changing world of exercises. However, I suggest that simply remembering a list of exercises limits your imagination and creative involvement in your process of exercising. Understanding the principles opens a world to create new, novel exercises that become exciting, enjoyable, and stimulate each individual mind and body. My hope is that you will "obey the principles without being bound by them" (Bruce Lee).

The chaotic exercise revolution driven by YouTubers is taking us away from the rigid protocols of our past military-style exercises. The novel *Tom Brown's School Days* by Thomas Hughes mirrors the exercises of the 1800s. Exercisers stood in regimental lines, religiously copying the master performing correct movements. The approach of the Victorian, Muscular Christianity movement was founded on moral principles of patriotic duty, discipline, self-sacrifice, and masculinity (a feminist version came later).

However virtuous the ideals, many of the methods of this one-size-fits-all approach also aimed to encourage a conformity of mind and body. One reason for this approach was the environment in which it existed: the emerging Industrial Revolution. The Industrial Revolution was a period in history dominated by machines, levers, pulleys, smoke, and toil. During this time, the working man was often perceived as little more than an extension of the machine, an idea brilliantly depicted by Charlie Chaplin in the 1936 film *Modern Times* (figure 5.1).

Conformity at the time of the machine was of paramount importance and, as such, robotic styles of movement were embedded by thought as well as deed. Humans often moved in physical and cognitive obedience like cogs in some large industrial machine that produced fodder for the First World War.

Figure 5.1. Charlie Chaplin in the film Modern Times*, 1936.*

Today the dominant machine is not one of cogs, gears, and hierarchical levels of importance. Present-day machines are ethereal mechanisms of inter-relationships better known as the chaos of the internet. During the Industrial Revolution, the language used to express understanding of human movement was dominated by concepts of levers and pulleys. Today it is fascia that seems to best mirror the chaotic, complex nature of our internet age. Entwined in our evolving web of movement appreciation is our knowledge of fascia. An understanding of our richest sensory organ—fascia—is starting to influence the world of exercise. In a world of fluidity, uncertainty, and changeability, it seems logical that our forms of movement and exercise should reflect our need for adaptability, resilience, and creativity.

If one's movement education has, to date, been limited to obediently repeating actions, this somewhat nonconventional approach may initially cause confusion and uncertainty. However, with time, the underlying implicit order emerges from the seemingly random and chaotic nature of principles-based learning. It is hoped that this principles-based approach allows the imaginative creation of new exercises and novel solutions to better train for the unpredictability of daily life.

A principles-based approach to movement allows the reader to explore movement rather than simply "doing" it. The exercises included in this book are examples only, not some kind of complete truth "as is" but a place from which to develop toward an ever-changing form of insight into self. The truth of movement then is a Kantian exploration rather than a certain thing to find (Kant, 1781). Movements start from within—one might say—from our core. The drive to move does not come from the extremities, the hand does not decide to reach out and lift a weight or grasp a cup. It is an inner thirst that drives the distal movement of the hand. It is the proximal center that initiates and controls distal movement, like the circles that radiate after a pebble is thrown into a pond.

Creating an inner space to sense the whole body in movement as a developing process is a concept at the heart of many practices such as yoga and martial arts. What is alien to masters of such practices is when movement is attempted as a passive "thing to be done" or the body as an object somehow "done to," devoid of emotion or meaning. Any exercises can become a dialogue between the self and the environment. Exercise can be an opportunity to build our relationship with space and time and with our inner and outer spaces. It is hoped that through playful and curious exercise there is an unfolding self-discovery that continues, not to some arbitrary final goal but toward increasing order and complexity. Exercise—in all its forms—allows us the space and time to engage in the total sensorium of the consciousness of movement.

Principles of exercise is not a new concept. Exercise books regularly give five, seven, or even ten principles of exercise. The most common include specificity, progression, overload, adaptation, and reversibility. It is hoped that further understanding can be gained by adding to and expanding upon some of these original principles of exercise.

Pain

The aim of the wise is not to secure pleasure, but to avoid pain. (Aristotle)

The utopian possibilities of physical and cognitive self-discovery associated with movement can come crashing down to earth by one concept—pain. It is paramount that pain is avoided to allow us to enjoy the myriad of benefits associated with movement. Exercises should not provoke pain. If pain does occur one must stop, it really is that simple.

Movement needs to be, and remain, a largely pleasurable act. While I advocate pleasurable, enjoyable, and pain-free movement, let us not get confused with ease. I am not advocating that all exercises should be easy, "Man needs difficulties; they are necessary for health" (Jung, in Miller, 2004).

At times, exercising can be hard, difficult, challenging, and evoke discomfort. Many exercisers will cry out in seeming pain—and when asked—will use colorful language to emphasize the "pain" of their chosen exercise. Most famously, rower Steve Redgrave declared "If anyone sees me go near a boat, you've got my permission to shoot me," after winning his fourth rowing gold medal in 1996, only to win his fifth gold in 2000.

This signifies that the pain of hard training is significantly different from other versions of pain. Sometimes it is hard to distinguish between pain and the normal discomfort of hard training. If in doubt, it is best to take the safe option and, rather than "pushing through" in an outdated, deluded, macho idea of "no pain, no gain"; when in pain, it is always wise to stop. Considering why that negative pain sense exists will help you to decide whether to stop the whole session or continue. Pain is a complex multifaceted concept that is well researched and better understood than ever before. In 2018 this was reflected by the International Association for the Study of Pain who formed a task force that took two years to define pain as:

> *An unpleasant sensory and emotional experience associated with, or resembling that associated with, actual or potential tissue damage. (Raja et al., 2020)*

There are various reasons why one should stop exercising in the presence of pain. For the confines of this book, we shall keep things simple. Pain is the body perceiving an actual—or the potential—threat of tissue damage. Feel the heat from a fire and you quickly move your hand out of the flames to avoid any potential tissue damage. This pain response is simple and obvious—yet, in some exercise settings, ignoring such a threat is encouraged. If you sense pain or discomfort, it is better to stop completely or modify the exercise rather than continuing and creating more problems.

There are various possible consequences of continuing to exercise despite one's body shouting stop. The most obvious is tissue damage to the area in pain. Less obvious—but perhaps more common—is the creation of poor actions and altered biomechanics. One strategy to avoid the painful area, perceived as under threat, is to move in suboptimal ways to avoid the sensitive area. While this strategy might be fine in the short term, in the long term this can create other issues. Poor body mechanics overburdens other parts of the system. For example, if your right knee is in pain during a squat and you modify this action without discovering why the knee is in pain, you may be shifting the workload to the hip that may, in time, dissipate its workload to the lower back and then to the neck. In clinics, like my own, it is very common to discover that the chronic pain in one area of the body is being caused by suboptimal biomechanics elsewhere in the system.

Listening to your body allows you to seek professional help before you start to ingrain poor movement patterns that may cause long-term issues for the future. The key, keep exercises fun, varied, and free of pain.

Specificity

Specificity is perhaps the most widely accepted principle of training, and yet, strangely, it seems it is the most often overlooked. Specificity indicates that training should closely reflect the activity one wishes to improve as part of the **SAID** (**S**pecific **A**daptation to **I**mposed **D**emands) principle. Given that this principle is so well accepted, it seems strange that exercises have needed to become weirder and more wonderful. Exercises seem to look less and less like the sport or activity that they are supposedly designed to help.

Recently the term *functional training* has become a useful wrinkle to the specificity principle. While the principles of functional training are sound, often nonfunctional exercises are included under this heading. The key to ensuring that exercises are functional and specific is that they look, smell, and feel familiar to the individual and to the activity (Tiberio, 2019). The example I gave—"Golf and Rotation," Chapter 3—of golfers performing strange pretzel-like yogic twists, is a classic example of a nonfunctional exercise for that activity. It is worth noting that for other activities and individuals, down dog with twist is a perfectly functional and specific exercise—it all depends upon the individual and the activity.

Movements Not Muscles

An underlying motive to write this book was to convey a sense of the unbroken wholeness and totality of the body.
I hesitated to write the section explaining muscles separately, yet my hope is that some understanding can be enhanced through such division. In movement, division often leads to suboptimal patterns. Experience has taught us that training individual muscles can create inefficient and even dysfunctional movement patterns.

Quantitative research supports the idea of training movements, not muscles, a qualitative fact long known by movers, dancers, and those attempting to represent movement in the arts. Many sculptors attempted to capture movement in stone and bronze, perhaps none better than Auguste Rodin (figure 5.2).

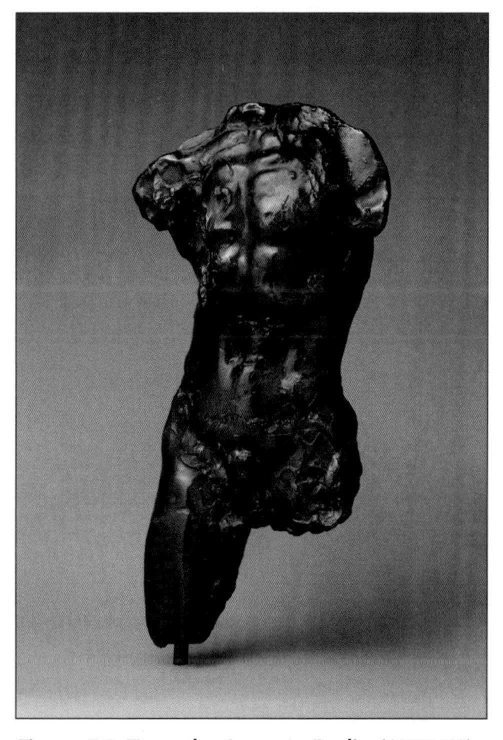

Figure 5.2. Torso *by Auguste Rodin (1877–78).*

During the process of sculpting, Rodin encouraged his models to be active, rather than holding a static pose. His initial sketches, that later became his sculptures, aimed to capture their attitudes in moments of movement. As Rodin explained, "Different parts of a sculpture represented at successive moments in time give an illusion of actual motion." His sculptures successfully "render inward emotion by muscular mobility" (Champigneulle & Brownjohn, 1980).

Within the clinical environment it is sometimes necessary to train individual muscles. However, it seems likely that this clinical intervention would not be necessary if the quality of whole-body patterns of movement had not, at some point, been neglected. It is thought that such "neglect occurs the minute we start to train partial movement patterns instead of whole movement patterns" (Cook, 2020).

Moving muscle-by-fragmented muscle, one will soon start to move as a stiff robot. Conscious actions are often awkward, inefficient, and likely to break down. Instead, consider the swimmer Michael Phelps who, as he enters the pool, seems to become liquid, or the elegant "attitude" of Carlos Acosta as he expresses the inexpressibility of music through dance, or the fluently repeating, easy speed as Mo Farah runs past without, it seems, touching the floor.

How these amazing performers gain such quality of movement is a hugely complex question. Like so many complex questions, humanity has attempted to break it down into manageable chunks, and so it is with movement. Our tradition has been to break down and teach movement and exercise part-by-part, muscle-by-muscle, and in this way, gradually progress. We now know this to be an inefficient and counterproductive method of learning to move. It would be fanciful to attempt to teach a toddler to walk in this same traditional manner. Imagine a two-year-old practicing first the arm swing, then hip extension, then pronation and supination of the feet, and then ask them to put it together to walk. Yet this inefficient, cognitively taxing, and complicated way of movement education is often used in the exercise setting.

Watching the developing gait of the toddler, the observer soon learns that movement patterns develop in stages of progressive difficulty. First the toddler's unsure steps move heavily along the edge of the sofa, then less solid support is found in the hands of a proud parent, developing to just holding one hand and soon achieving the definition of being human: becoming a "featherless biped" (Plato).

In time, the toddler gradually progresses to more complex environments until they are free, relaxed, and moving at speed over any surface, changing direction at a whim, and scaring adults as they race through parks and playgrounds. Training that attempts to reflect this developmental style should be applauded. Such developmental training gradually adds complexity and so variety to each successful movement. The emphasis here is the gradual development of movement literacy through variety.

Just as the developing toddler needs the strength, mobility, and stability to initiate movement, so exercises develop from proximal to distal, from deep to superficial. In this way, exercisers begin with low load, slow, predictable movements and develop toward higher load, dynamic, resisted, unpredictable variations. At each step of one's development, a priority should be given toward the quality of movement and maintain the specificity to the overall movement goal.

Quality is never an accident. It is always the result of intelligent effort. (John Ruskin)

Recognizing the quality of movement is a subjective skill that we all have. Every four years, many of us become experts in a sport without any education or previous knowledge. During the Olympic Games, millions watch and soon become adept at deciding which of the synchronized divers, skateboarders, or equestrian riders will win a medal. While it is advisable to have a professional to observe and critique your exercises, a willing friend will often help, and failing that, use your phone.

In the past, only the privileged few had access to video analysis—today many of us have that resource. Watching and rewatching yourself exercise can help you to identify some of the idiosyncrasies and difficulties, and then you can start to improve. You might notice jerky actions that lack the flow and efficiency expected. Or you might observe the moment when you go "off center" or cannot maintain an expected position. You might spot other aspects of interest—such as when a joint moves in a strange or unexpected way—indicating suboptimal biomechanics. Each observation allows the possibility for adjusting the technique. It is often a good idea to reduce the forces (such as reducing the weight) until more optimal movements can be achieved.

Analysis by video, friend, or expert allows you to recognize which phase of motor learning you are at. The classic theory of motor learning by Fitts and Posner (Fitts & Posner, 1967) tells of three stages of motor learning. Other theories exist but this is still considered a useful starting point. Identifying which stage you are at will help you decide when you are ready for progression. As the name suggests, at the first—or cognitive—stage, movements are carefully thought through. At this early stage, one is heavily reliant on feedback and instructions. This is the phase where video analysis is a particularly useful method to self-check the gross actions when many errors are likely to occur.

With time and dedication, one moves into the second—or associative—phase. With fewer mistakes and more consistency, the cognitive demands diminish. Movements are not yet automatic but require concentration to be conscious competent.

With significant practice, movements become automatic—phase three—when movements become unconscious and habitual. Very little thinking occurs as movements have become "second nature" and performed easily. At this stage, further demands can be made such as transferring the same task to different environments (Fitts & Posner, 1967).

Neuroplasticity

> *Any man could, if he were so inclined, be the sculptor of his own brain. (Cajal, 1897)*

While physical developments are necessary to progress through the phases, so too are changes in the brain. Neuroplasticity refers to "the ability of the nervous system to respond to intrinsic and extrinsic stimuli by reorganizing its structure, function, and connections" (Raja et al., 2020).

A little appreciation of the theory of neuroplasticity allows the performer to improve their skill development. At the cognitive phase of learning, listening to music and other distractions reduces our capacity to develop the awareness needed at this stage. A better learning environment is quiet, which allows an inner focus. Training with meaning allows purposeful exercises, further enhancing one's neuroplastic development. It is hoped that by understanding the *why* of training, this book has gone someway to make training more meaningful. While I advocate variety above repetition in much of this book, I also acknowledge the need for massed practice. Repetition allows the fine-tuning necessary for better timing and recruitment of muscles and precision and efficiency of movement (Tsao & Hodges, 2007). Add plenty of positive feedback, and the development from novice to expert can rapidly progress.

Repetition—coupled with a global movement perspective—improves an important component of movement: timing. It has been found that "muscle activity over the spine changes instantaneously in response to the loads applied and the tasks required" (Vleeming et al., 2007). Isolated training of individual muscles is now thought to be either very difficult, doubtful, or impossible. Improving the quality of movements targets multiple muscles and allows improvement of the important timing and coordination.

> *Lack of sufficient coordination in core musculature can lead to decreased efficiency of movement and compensatory patterns, causing strain and overuse injuries. Thus, motor relearning of inhibited muscles may be more important than strengthening in patients with LBP [lower back pain] and other musculoskeletal injuries (Akuthota et al., 2008).*

Figure 5.3 is an example of an exercise to promote coordination and timing for runners. This exercise creates a low-load situation, and the small bounces on the ball create a pulsing vertical stimulus similar to that experienced when running. In this exercise, the runner must deal with the vertical forces of gravity and ground reaction force through the torso. The movement stimulus encourages a neuromuscular response to each pulse, encouraging timing, preparation, and appropriate muscle behavior. This exercise has the added benefit of training without further work to a runner's tired legs, and as a stationary exercise, it allows a competent friend or coach to observe any losses in the biomechanical integrity of the torso in dealing with these forces. I often use this exercise in clinic and am now able to identify precise energy leaks that can appear to occur at one of the costotransverse joints.

Developing variety, complexity, and demand is dependent upon the success of the individual at each stage (figure 5.4).

Figure 5.3. Rhythmical bouncing while performing running arms to help timing and coordination.

Figure 5.4. Rhythmical bouncing while performing running arms and holding one dumbbell, which increases difficulty and rotational work.

Sports- and Non-Sports-Specific Exercises

> *I hate sports the way people who like sports hate common sense.*
> *(H. L. Mencken)*

The principles outlined here are relevant to enhance the training for everyone's body, goal, and lifestyle demands. Of course, not everyone exercises to improve their athletic performance. For non-sports people, as with sports people, exercises must be specific to—and focused upon—the demands of life. Many everyday activities are of low load (light weights and low forces) and slower than the athletic extremes, and exercises should be representative of these differences. The gym represents a safe, consistent environment in which to explore and train for the inconsistent and uncertain real world. Exercises that enhance one's ability to dig in the garden, squat down into a chair, or carry shopping, help prevent injury and increase one's enjoyment of everyday tasks.

Sports seems to be an experiment into the extremes of human capabilities. Exercises reflect the extreme nature of many sports by using higher loads and greater resistance and take place in an unpredictable dynamic environment of change. While the temptation is to get overly complex, training should remain specific to the sport.

To further understand the principle of specificity (page 95), let us consider an exercise synonymous with the core and stability, the plank (see figure 1.7) and its little brother, the side plank (figure 5.5).

As we will find, these exercises are useful for some tasks yet possibly detrimental to other tasks. The plank and the side plank certainly work many of the abdominal muscles we have been discussing throughout this text. Both planks—and other variations—are regularly described as exercises to improve stability. For this reason, one might conclude that the plank would help an athlete who is in an unstable environment.

Figure 5.5. The side plank exercise.

Sitting in a rowing scull is one of the most unstable situations for an athlete. Many rowers have spent optimistic hours holding a plank in the hope that this static position will transfer to improve the dynamic nature of rowing. The plank exercise teaches the body to be static, solid, and not to move. Sit in a rowing scull, or any narrow boat, and one quickly realizes the need to respond rapidly to the movements of the boat. The tilt of the boat requires the pelvis to tilt while maintaining an upright torso. In this way, the pelvis and torso must disassociate, the spine must bend to allow the forces to be dissipated, and the movement of the boat controlled. This is quite the opposite of a plank that trains the torso and pelvis to be held together. Hold the torso and pelvis in a solid plank of neutrality, and the tilt of the boat—that tilts the pelvis—would tilt the solid torso and thus cause the rower to become a swimmer!

An example of a more functionally specific exercise can be found in figure 5.6. An alternative exercise uses an over-ball (a partially deflated ball designed to sit on) to promote, allow, and train the capacity of the athlete to tilt the pelvis independently from the torso in a controlled manner (figure 5.7). The exercise mimics the unstable platform of the boat and is specific to training athletes such as rowers and kayakers.

Figure 5.6. A functionally specific version of the sit-up mirrors a rowing action.

Figure 5.7. Sit-ups on an over-ball increases instability. Useful for athletes in unstable environments such as kayakers and rowers.

With a little modification, this exercise can become useful for office workers to encourage dynamism to an otherwise static sitting position. Using a larger (Swiss) ball to sit on allows both feet to remain on the floor and mimics sitting at the desk. A slow, controlled tilting of the pelvis is encouraged to promote mobility; taking one foot off the ground trains the dynamic strength and balance necessary for sitting. Further modifications allow this exercise to mimic walking, an action that requires a controlled pelvic tilt without excessive side bending of the lumbars. For further information on this, I refer you back to the discussion relating to figures 3.26 and 3.27.

Variety

> *Nothing is pleasant that is not spiced with variety. (Francis Bacon)*

The multidirectional warp and weft of the muscular and fascial matrix tells of a need for variety. Varying one's patterns of movement better enables us to deal with the multiple directions of force in everyday living. Varying the environment means a change in the direction, tempo, surface, load, position, or duration of exercises.

Minimizing the boredom of repeatedly doing the same action reduces the physical wear and tear on the body. For example, the monotony of running on a treadmill seems to typify this in comparison with running outside. The floor of the treadmill is consistent and supports our flexible and dynamically intelligent feet. Conversely, running outside forces the feet—and whole body—to adapt to different terrains, surfaces, and gradients. Variety keeps movement fun; joyful movement enhances motivation and continues to inspire. The mental health benefits of joyous movement are well documented (Dunn & Jewell, 2010; Martin, 1977; Mikkelsen et al., 2017) and observable in children's playparks and continue into adulthood through sports and exercise.

It is all very well saying one should have variety in one's training, but without a framework, the search for variety can be random and chaotic. One difficulty in constantly looking for new exercises is one often finds similar, habitually similar versions of our favorite exercises, even when we know the importance of a varied diet of exercises. The framework to build variety into movement that I suggest is based upon the work of the Gray Institute, which uses the three planes of movement. Each plane—sagittal, frontal, and transverse—gives rise to specific moment directions, as well as positional indicators.

CHAPTER 6

Body Planes

Sagittal Plane

In the body, the sagittal plane divides the body into a right and a left side. Movements within the body in the sagittal plane include flexion and extension, as occurs in the lumbar spine. Move the whole body in the sagittal plane, and one moves in either an anterior (forward) or posterior (backward) direction. A forward or anterior lunge—actively stepping forward—is an example of the whole person moving in the sagittal plane. A step back or posterior lunge is another example of a sagittal plane movement. Starting in a lunge position is an example of pre-positioning oneself in the sagittal plane, with one foot forward and the other back. As we shall discover from this sagittal plane starting position, one can add variety by using the other two planes. Sagittal movements are not confined to the feet and lunges. By reaching both hands out in front or behind, you are also moving in the sagittal plane.

Frontal Plane

The frontal plane divides the body into a front and a back. Movements within the body in the frontal plane include side bends of the torso and abduction and adduction of the limbs. An example of moving the whole body in the frontal plane includes taking step to one side—a lateral lunge. In this lateral lunge example, both legs move in the frontal plane, one by actively moving, the other passively responding.

Wave your hands in a large arch over your head, as you might do at a rock concert (figure 6.1), and you are largely moving in the frontal plane—arms are waving in adduction and abduction and back again as the torso sidebends to the rhythmical strum of the bass guitar. Combine the two and perform a frontal plane cartwheel. Pre-positioning one's feet in the frontal plane starts to add a little variety. Exercises that may traditionally have started with feet hip-width apart can now be explored from a wider or narrower stance.

Figure 6.1. Wave your hands in the air like you just don't care!

Transverse Plane

The transverse plane divides the body into upper (superior) and lower (inferior) halves. Movements within the transverse plane are rotational in nature—these include the internal and external rotation of the limbs and rotation of the torso—examples include walking like a duck (feet and legs in external rotation) or with "pigeon toes" (feet and legs in internal rotation).

Turning to see who is behind you is a good example of the body moving in the transverse plane. First one might turn one's head (a transverse plane rotation) but can't quite see, so one turns the torso, then twists right around, and the pelvis and legs start to move (yes, you've guessed it) in the transverse plane. Continue this movement and one might step around, in a transverse plane rotation.

Reaching out in front of you with one hand is a movement that starts in the sagittal plane. Reach further and you'll soon notice your torso rotating. Increase the rotational component by reaching diagonally across to the opposite side. By reaching across your body in this manner, your arm has adducted: a simple movement that demonstrates the interconnecting aspects of the planes of movement.

A Combination of Planes

Education on planes of movement often stops after the explanation of each plane. Where it becomes useful and more applicable to everyday movements is by combining planes. Systematically working through each plane can give a huge variety to even the simplest exercises. As an example—which I hope becomes a catalyst to exploration—let us consider a simple and common exercise: the squat.

Taking a slightly more detailed biomechanical view of a standard, traditional squat will, I hope, help identify energy leaks or areas that might need attention. The expected biomechanics of the squat should include—on the way down—one's feet pronating. The whole foot widens around the axis of the second ray (second metatarsal). The calcaneus and talus should both medially tilt, and the foot should go into dorsiflexion. The knee should flex, with a slight medial rotation of the tibia relative to the femur during the first 15% of knee flexion (Goodfellow et al., 1976). The hip should flex, and the pelvis should remain neutral (no torsions or rotations) but tilt anteriorly relative to the femur. The torso should remain neutral relative to the pelvis. The head and neck also remain neutral or posteriorly tilted to keep the eyes on the horizon. The shoulders and arms mirror the activity as needed. Reverse all this to describe the actions of coming up from a squat.

Now let's start to play with the planes of movement in a systematic manner to show some of the varieties possible for the squat. By giving a few examples, it is hoped that the reader will be able to work out all the variations possible. I shall not use up pages and pages to describe all the possibilities because once the principles are understood, the gym becomes your playground of exploration. Figuring out the various possibilities of different planes is also a great way to train your brain and improve the cognitive resilience necessary as we age (Horowitz et al., 2020).

For example, let us consider the feet position during a squat. In the standard squat, feet are positioned hip-width apart and face forward. Playing in the frontal plane allows a wider or narrow foot position for the squat. In the sagittal plane, we can pre-position our feet forward or back, much like a lunge. Internal or external rotation allows further work in the transverse plane. Combining these planes allows a wide squat (frontal plane variation), with the right foot externally rotated (transverse plane variation), and the left foot forward (sagittal plane variation).

The squat variations demonstrated in this text use the front-loaded barbell position. Change the equipment to dumbbells and the variation of arm positions can be used using the same principle. Now, for example, the right arm is anterior and narrow, while the left is wide and exteriorly rotated as the squat is performed. This strange squat-like action is certainly unconventional but, as discussed, its benefits are numerous.

Many of the photographs show how much our models had to work during the photoshoot, indicating that these positions challenge even the fittest squat experts. A variety of squats is presented in figure 6.2.

This principle of creating variety can now be taken into almost any exercise. It allows variety in conventional exercises like the push-up (figure 6.3) and sit-up (figure 6.4), and allows variety in playing with functional training for sports such as golf (figures 3.22–3.23) and running (figure 6.5). As you can see, this principle has been used throughout this book to increase the forces and workloads to specific areas.

Neutral or hip-width in frontal, neutral in sagittal, and neutral in transverse

Narrow in frontal, neutral in sagittal, and neutral in transverse

Wide in frontal, neutral in sagittal, and neutral in transverse

Neutral in frontal, neutral in sagittal, but right foot toed-out in transverse

Figure 6.2. Variety of squats.

Figure 6.2 illustrates a series of squats showing some of the possible combined variations of foot positions in each plane. While not all variations are shown, for example, wide in frontal plane, right foot anterior and left foot posterior in the sagittal plane, right foot toed-in and left foot toed-out in the transverse plane, this

Neutral in frontal, neutral in sagittal, but left foot toed-in in transverse

Narrow in frontal, neutral in sagittal, but right foot toed-in in transverse

Narrow in frontal, neutral in sagittal, but right foot toed-out in transverse

Narrow in frontal, neutral in sagittal, but right and left feet toed-in in transverse

Figure 6.2. (continued)

gives a comprehensive set of examples of the principles that can be used with various other exercises. It is worth exploring all variations.

Playing with these plane-based variations creates sixty-three possible squat combinations. More if you included a crossed over foot position, and many

Narrow in frontal, neutral in sagittal, but right and left feet toed-out in transverse

Wide in frontal, neutral in sagittal, but right foot toed-out in transverse

Wide in frontal, neutral in sagittal, but right foot toed-in in transverse

Wide in frontal, neutral in sagittal, but right and left feet toed-in in transverse

Figure 6.2. (continued)

more if you were to add varying degrees of toed-in, toed-out, internal and external rotation, and varying degrees of wider or narrower stance.

Going through all these variations is a significant workout for your memory as well as your body. By adding frontal, transverse, and sagittal arm positions

Wide in frontal, neutral in sagittal, but right and left feet toed-out in transverse

Neutral in frontal, left foot anterior and right foot posterior in sagittal, but neutral in transverse

Neutral in frontal, left foot anterior and right foot posterior in sagittal, but left foot toed-in in transverse

Neutral in frontal, left foot anterior and right foot posterior in sagittal, but left foot toed-out in transverse

Figure 6.2. (continued)

and movements the variations are (almost) endless. As you play with these movements, you will recognize how you can tailor your exercises to be more functionally specific to your daily tasks and goals.

Neutral in frontal, left foot anterior and right foot posterior in sagittal, but right foot toed-out in transverse

Neutral in frontal, left foot anterior and right foot posterior in sagittal, but right foot toed-in in transverse

Neutral in frontal, left foot anterior and right foot posterior in sagittal, but right and left foot toed-in in transverse

Neutral in frontal, left foot anterior and right foot posterior in sagittal, but right and left foot toed-out in transverse

Figure 6.2. (continued)

Figure 6.3. Variety of push-ups. Some push-up variations applying the same principle used to vary the foot positions for the squat, figure 6.2, but this time for the push-ups. Not all variations are shown but the reader is encouraged to explore all possibilities.

Figure 6.4. A series of sit-ups with arm movements to show some of the possible variations. Adding a weight increases the work to decelerate and can be used in all three planes of movement.

It is now possible to use this same principle for the imaginative use of pieces of equipment to change the load and task demands. For example, elastic resistance bands pulled at different planes challenge and work the body to remain in position. Adding a weight increases the decelerating capacity of the movement. Further examples for the sagittal plane (figure 6.6), frontal plane (figure 6.7), and transverse plane (figure 6.8) are included, with many more from your imagination.

Adding variety to the simplest or more complex activities and exercises allows one to target all fibers of all muscles in all directions. It allows muscles to work in both their concentric and eccentric actions to accelerate or decelerate the various forces acting on them. Such exploration quickly identifies movements that are more difficult for each individual, possibly highlighting areas of relative weakness that need further training.

Critically, such exploration adds fun to otherwise mundane repetitive exercises. Attempting to dribble a weighted basketball while performing a one-legged squat challenges basketball and non-basketball players alike (figure 6.9). The basketball adds a level of unpredictability, increasing the dynamism, balance, and control to this physical and cognitively challenging activity. Various tweaks could easily be made to make it more specific to an individual, their goals, and tasks.

Figure 6.5. A suspension trainer allows the athlete to increase the angle of lean and mimic the actions of running. Alternating the contralateral arm-drive and/or single-leg drive the athlete actively pushes down into the ground to engage the whole system. Maintaining this single-leg loaded position forces the athlete to control pelvic torsion and thoracic rotation. This creates forces like those experienced during running while improving aspects of coordination, balance, and timing into running-specific training.

Figure 6.6. An example of a sagittal plane variation with elastic band to add to workload.

Figure 6.7. An example of a frontal plane variation with elastic band to add to workload and the muscles actively decelerating arm action.

Figure 6.8. An example of a transverse plane variation with weight to add to workload.

Figure 6.9. Dribbling a basketball while performing one-legged squats for fun and dynamic dexterity.

CHAPTER 7

Pulling it All Together

Prevention

It is easier to prevent bad habits than to break them. (Benjamin Franklin)

Most of my working life has been spent in clinics dealing with a variety of injuries and complaints. From this experience, I see the importance of movement with a preventive theme. Unfortunately, using exercise as a prevention measure is still rare—at least in the West—whereas in China, "Tai chi is the predominant form of exercise" for its health benefits (Birdee et al., 2013). It is farcical to expect that two gym sessions a week will offset an otherwise sedentary existence.

A lifestyle with multiple pockets of movement is clearly much closer to the ideal. One striking image from the paleoanthropologist Daniel Lieberman's book *Exercised* (Lieberman, 2021) is of hunter-gathers sitting around a fire. Surprisingly, various researchers, like Lieberman, indicate that hunter-gathers sit for as long as office workers. The big difference is not the overall time but that the hunter-gather gets up at regular intervals to tend the fire, deal with the children, stir the dinner, and other periods of brief activity. This is contrasted with statuesque TV watchers waiting for their food to be delivered.

In promoting movement, the exercise professional could learn a lot from dentists. Somehow dentists have convinced us all that teeth brushing is a normal, habitual part of the daily routine. Like brushing teeth, some exercises may not be much fun. Who brushes one's teeth for enjoyment? We all brush to prevent tooth decay; why don't we all exercise to prevent back pain, incontinence, and depression (to name just a few well-documented benefits of movement)? The utopian dream is a world where exercise is a normal, expected part of maintaining good health and clinics like mine are empty. In such an idealistic world, exercises are not performed in hindsight, nor are they reactionary to a problem, but they are preventive and hopeful.

Further Exercise Examples

The following exercises are included to act as a catalyst to provoke imaginative movement exploration. Couple these examples with the principles outlined earlier in this book to create a large repertoire of functionally relevant exercises that are specific to your life, your sport, and your goals.

Figure 7.1. A kettlebell swing is an excellent way to train the capacity of various muscles to decelerate a force. In this example, various muscles, including tibialis anterior, rectus femoris, rectus abdominis, sternalis, and latissimus dorsi are coordinating to lengthen under tension to control the swing of the kettlebell. Exploring the various foot positions, as seen in figure 6.2, adds further variety and focus.

Figure 7.2. Single-armed kettle bell swing increases the rotational forces. Muscles, including the hamstrings, erector spinae, and quadratus lumborum work to resist the downward swing of the kettlebell.

Figure 7.3. Here the athlete has added an elastic resistance band to a conventional push-up. The band creates a lateral/diagonal force that the athlete must resist to maintain a neutral pelvis-to-torso relationship. In this example, the left arm applies an increased lateral force and the right arm an increased medial force to continue the push-up action.

Figure 7.4. Here a red elastic band has been added to a conventional goblin squat. The band adds additional work to the abductors, adductors, obliques, and many other muscles to resist rotating and maintain an upright position.

Figure 7.5. The dumbbell curl is not only for the biceps. Holding an elastic band in the opposite hand increases the workload of differing lines of obliques and horizontal fibers across the torso as the dumbbell is raised and lowered.

Figure 7.6. Playing catch while in a plank position adds fun, competition, and dynamism to an otherwise static exercise. The occasional one-armed position and unpredictability of your partner's throw increases work to resist rotation.

Figure 7.7. This progressively challenging exercise promotes awareness and motor control. (a) The athlete is asked to consider the perceived weight of the left leg compared to the right leg, which may indicate poor muscle recruitment; (b) Using her hands, the athlete monitors for any unwanted rotations as she rolls up into the position in the photograph and rolls back down, vertebra by individual vertebra, without rotating; (c) Lifting and extending one leg without rotating or dropping the pelvis on one side is a challenge for many athletes.

Spinal Mobility Exercises

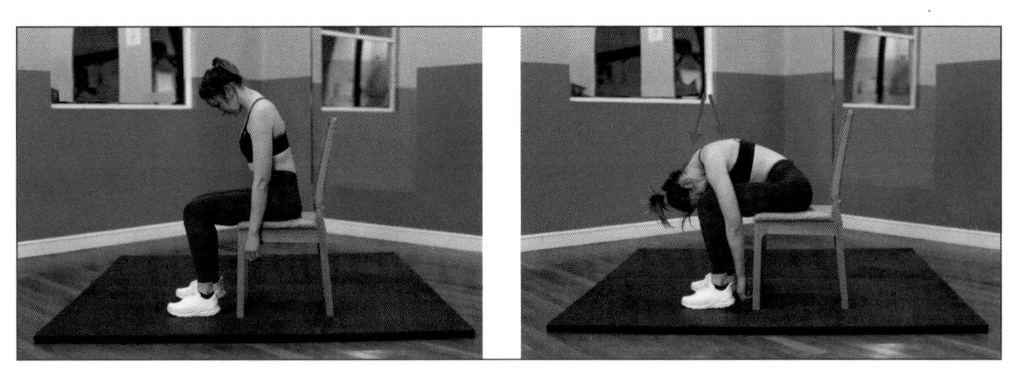

Figure 7.8. Seated roll-down for spinal mobility. Rolling down and back up by moving one vertebra at a time helps spinal mobility. The aim is to have an even, rounded curve. This athlete has a flat area that has failed to flex (see arrows). This indicates an area that needs more work to allow this athlete to better dissipate the forces of flexion.

Figure 7.9. Creating a stretch to the quadratus lumborum by prepositioning knees and hands as shown. The athlete then takes a breath and shifts the pelvis back, as indicated by the arrow.

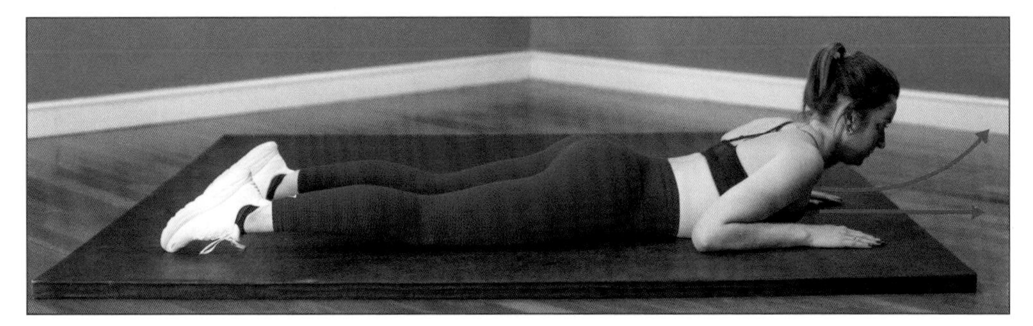

Figure 7.10. The mini cobra (or baby backbend) helps promote thoracic extension. Starting flat with forehead on the ground, the athlete extends to lift vertebra by vertebra and "peel" the sternum off the floor.

Figure 7.11. Seated thoracic mobilization in all planes. (a) In the sagittal plane reaching forward. A useful cue is to move from the fingertips and "pulse" for a few repetitions; (b) A frontal plane large waving action, left to right; (c) In the transverse plane, a diagonal reach and again pulsing action to hydrate and mobilize the associated tissues; (d) Similar to (c) and also in the transverse plane, a diagonal reach backward; (e) By combining the transverse plane diagonal reach with the frontal plane "wave" we can utilize our understanding of spinal mechanics to gap and mobilize the joints of the upper thorax.

Figure 7.12. Offset loaded lunge.

Figure 7.13. Offset loaded squat.

During an overhead lunge we loaded only one side of the barbell. In the photograph, the right foot is pushing down as the right arm is push upward and the left arm is pulling down to keep the barbell level. The torso works to resist the thoracic rotation and pelvic torsion created as the left leg flexes. The adductors work to control the hip from falling into abduction on the weighted side.

A further variation on the squats shown in figure 6.2 is to load only one side of the barbell. The latissimus dorsi on the non-weighted side must engage to pull down as the weighted side resists the downward weight to maintain horizontal balance. The obliques work to resist the contralateral torsion of forces created across the torso. While the increased workload on the weighted side is obvious, the non-weighted leg must also work to control and balance the whole system.

Caveat to the Exercises Shown

I hope the reader will not be weighed down by the theoretical information of the last few pages. Allow the theory to inform while allowing the exercises to be a catalyst for ideas. I hope this book will provide support and stimulate ideas for the development and exploration of complex, varied—and above all—enjoyable movements.

The exercises are not meant to be a perfect or complete list. It is worth noting that the pictures of the exercises were not chosen based on gender or body type. While every effort was made to get the exercises—and so the pictures—"correct," they invariably show the idiosyncratic nature of the individuals shown. It was fate that gave a more mobile female and solid male physique to be used in the pictures, and nothing more is meant by this. Some of the pictures show the difficulties, struggles, and challenges both had when asked to perform some of these exercises. These difficulties emphasize the importance of gauging the developmental stage of training to be specific to the individual and the demands of their life. For example, to demonstrate the positive benefits of offset loaded tasks, both our models had to significantly reduce the weight they might normally use for even lifts.

Lastly, I hope the pictures convey the humor and fun we all had in taking these pictures—it is hoped that humor, fun, and enjoyment is at the heart of your movement.

> *In my end is my beginning. (Eliot, 1941)*

References

Akuthota, V., Ferreiro, A., Moore, T., & Fredericson, M. (2008). "Core stability exercise principles." *Current Sports Medicine Reports*, *7*(1), 39–44.

Akuthota, V., & Nadler, S. (2004). "Core strengthening." *Archives of Physical Medicine and Rehabilitation*, *85*(1), 86–92.

Archimedes. (287 BCE). *The Works of Archimedes* (Heath Thomas, Ed.). Dover Publications, 2002.

Barral, J. P., & Mercier, P. (1988). *Visceral Manipulation*. Eastland Press; Revised edition 2006.

Beinfield, H. (1992). *Between Heaven and Earth: A Guide to Chinese Medicine*. Ballantine Books Inc.; 1st Trade Pbk. edition.

Bergmark, A. (1989). "Stability of the lumbar spine. A study in mechanical engineering." *Acta Orthopaedica Scandinavica. Supplementum*, *230*, 1–54.

Birdee, G. S., Cai, H., Xiang, Y. B., Yang, G., Li, H., Gao, Y., Zheng, W., & Shu, X. O. (2013). "T'ai chi as exercise among middle-aged and elderly Chinese in urban China." *Journal of Alternative and Complementary Medicine (New York, N.Y.)*, *19*(6), 550–557.

Bohm, D. (1980). *Wholeness and the Implicate Order*. Routledge Classics 2002.

Bojairami, I. El, & Driscoll, M. (2022). "Coordination between trunk muscles, thoracolumbar fascia, and intra-abdominal pressure toward static spine stability." *Spine*, *47*(9), E423–E431.

Brown, S. H. M., & McGill, S. M. (2008). "An ultrasound investigation into the morphology of the human abdominal wall uncovers complex deformation patterns during contraction." *European Journal of Applied Physiology*, *104*(6), 1021–1030.

Bump, R. C., Hurt, W. G., Fantl, J. A., & Wyman, J. F. (1991). "Assessment of Kegel pelvic muscle exercise performance after brief verbal instruction." *American Journal of Obstetrics and Gynecology, 165*(2), 322–327; discussion 327–329.

Buttaccio, J. (2018, January). "Kegels for Men are a Thing, and You Should Absolutely Be Doing Them." *Men's Health Magazine.*

Cajal, R. Y. S. (1897). *Advice for a Young Investigator.* MIT Press 2004.

Canadian Lung Association. (2022). https://www.lung.ca/lung-health/lung-info/breathing.

Capra, F. (1992). *The Tao of Physics: An Exploration of the Parallels Between Modern Physics and Eastern Mysticism.* HarperCollins Publishers.

Chaitow, L., Gilbert, C., & Bradley, D. (2013). *Recognizing and Treating Breathing Disorders: A Multidisciplinary Approach* (2nd ed.). Churchill Livingstone.

Champigneulle, B., & Brownjohn, M. J. (1980). *Rodin (World of Art).* Thames and Hudson Ltd.

Cohen, K. S. (1999). *The Way of Qigong: The Art and Science of Chinese Energy Healing.* Ballantine Books Inc.

Cook, G. (2020). *Movement: Functional Movement Systems: Screening, Assessment, Corrective Strategies.* On Target Publications.

Dillner, L. (2018, March 5). "Are you sitting comfortably: the myth of good posture." *The Guardian.* https://www.theguardian.com/lifeandstyle/2018/mar/05/are-you-sitting-comfortably-the-myth-of-good-posture.

Dufour, S., Vandyken, B., Forget, M. J., & Vandyken, C. (2018). "Association between lumbopelvic pain and pelvic floor dysfunction in women: A cross sectional study." *Musculoskeletal Science & Practice, 34,* 47–53.

Dunn, A. L., & Jewell, J. S. (2010). "The effect of exercise on mental health." *Current Sports Medicine Reports, 9*(4), 202–207.

Earls, J. (2020). *Born to Walk: Myofascial Efficiency and the Body in Movement* (2nd ed.). Lotus Publishing.

Earls, J., & Myers, T. (2017). *Fascial Release for Structural Balance* (revised ed.). Lotus Publishing.

Eliot, T. S. (1941). *Four Quartets.* Faber & Faber; Main edition 2001.

Ellenbecker, T. S., Pluim, B., Vivier, S., & Sniteman, C. (2009). "Common injuries in tennis players: Exercises to address muscular imbalances and reduce injury risk." *Strength and Conditioning Journal, 31*(4), 50–58.

Encyclopedia Britannica online. (2022). https://www.britannica.com/Science/Life.

Engel, G. L. (1977). "The need for a new medical model: A challenge for biomedicine." *Science (New York, N.Y.), 196*(4286), 129–136.

Epstein, D. (2020). *Range: How Generalists Triumph in a Specialized World.* Macmillan.

Faries, M., & Greenwood, M. (2007). "Core training: Stabilising the confusion." *Strength and Conditioning Journal, 29*(2), 10–25.

Fitts, P. M., & Posner, M. I. (1967). *Human Performance*. Greenwood Press 1979.

Fuller, B. R., & Applewhite, E. J. (1982). *Synergetics: Explorations in the Geometry of Thinking*. Prentice Hall & IBD; New edition.

Gilman, S. L. (2018). *Stand Up Straight!: A History of Posture* (1st ed.). Reaktion Books.

Gleick, J. (1997). *Chaos*. Vintage; Reprint edition.

Goodfellow, J., Hungerford, D. S., & Zindel, M. (1976). "Patello-femoral joint mechanics and pathology. 1. Functional anatomy of the patello-femoral joint." *The Journal of Bone and Joint Surgery. British Volume*, *58*(3), 287–290.

Gorman, D. A. (1981). *The Body Moveable* (6th 2014). LearningMethods Publications.

Gracovetsky, S. (1988). *The Spinal Engine*. Lulu.com.

Gros, F. (2008). *A Philosophy of Walking*. Verso Books; Reprint edition.

Guimberteau, J. C., & Armstrong, C. (2015). *Architecture of Human Living Fascia: The Extracellular Matrix and Cells Revealed Through Endoscopy*. Handspring Publishing Limited.

Hakim, A., Keer, R., & Grahame, R. (2010). *Hypermobility, Fibromyalgia and Chronic Pain* (A. J. Hakim, R. J. Keer, & R. Grahame, Eds.). Churchill Livingstone.

Han, J. N., Stegen, K., de Valck, C., Clément, J., & van de Woestijne, K. P. (1996). "Influence of breathing therapy on complaints, anxiety and breathing pattern in patients with hyperventilation syndrome and anxiety disorders." *Journal of Psychosomatic Research*, *41*(5), 481–493.

Hawkins, M. (2016). "The diaphragm—not just for breathing." InMotion Spine & Joint Center. https://www.imsjc.com/articles/the-diaphragm-not-just-for-breathing.

Henderson, D. (2008). *'Scuse Me While I Kiss the Sky: Jimi Hendrix: Voodoo Child*. Atria Books.

Herrington, L. (2011). "Assessment of the degree of pelvic tilt within a normal asymptomatic population." *Manual Therapy*, *16*(6), 646–648.

Hibbs, A. E., Thompson, K. G., French, D., Wrigley, A., & Spears, I. (2008). "Optimizing performance by improving core stability and core strength." *Sports Medicine (Auckland, N.Z.)*, *38*(12), 995–1008.

Hippocrates. *Hippocratic Writings*. Chicago: Encyclopedia Britannica, 1955.

Hodges, P., Cholewicki, J., & van Dieen, J. H. (2013). *Spinal Control: The Rehabilitation of Back Pain: State of the Art and Science* (1st ed.). Churchill Livingstone.

Horowitz, A. M., Fan, X., Bieri, G., Smith, L. K., Sanchez-Diaz, C. I., Schroer, A. B., Gontier, G., Casaletto, K. B., Kramer, J. H., Williams, K. E., & Villeda, S. A. (2020). "Blood factors transfer beneficial effects of exercise on neurogenesis and cognition to the aged brain." *Science (New York, N.Y.), 369*(6500), 167–173.

Jones, M., & Rivett, D. (2019). *Clinical Reasoning in Musculoskeletal Practice* (2nd ed.). Elsevier.

Joyce, J. (1914). *Dubliners*. Penguin Classics 2000.

Kahneman, D. (2012). *Thinking, Fast and Slow* (1st ed.). Penguin.

Kant, I. (1781). *Critique of Pure Reason*. Penguin Classics 2007.

Korakakis, V., O'Sullivan, K., O'Sullivan, P. B., Evagelinou, V., Sotiralis, Y., Sideris, A., Sakellariou, K., Karanasios, S., & Giakas, G. (2019). "Physiotherapist perceptions of optimal sitting and standing posture." *Musculoskeletal Science & Practice, 39*, 24–31.

Lasseter, J. (1995). *Toy Story*. Pixar Animation Studios, Walt Disney Pictures.

Lederman, E. (2010). "The myth of core stability." *Journal of Bodywork and Movement Therapies, 14*(1), 84–98.

Lee, D. (2018). *The Thorax: An Integrated Approach* (1st ed.). Handspring Publishing.

Lee, D. G. (2010). *The Pelvic Girdle: An Integration of Clinical Expertise and Research* (4th ed.). Churchill Livingstone.

Lee, D. G., Lee, L. J., & McLaughlin, L. (2008). "Stability, continence and breathing: The role of fascia following pregnancy and delivery." *Journal of Bodywork and Movement Therapies, 12*(4), 333–348.

Levine, J. A. (2014). *Get Up!: Why Your Chair is Killing You and What You Can Do About it*. Griffin.

Liao, W. (2001). *T'ai Chi Classics*. Shambhala Publications Inc.

Lieberman, D. (2021). *Exercised. The Science of Physical Activity, Rest and Health*. Penguin.

Liem, T., Tozzi, P., & Chila, A. (2017). *Fascia in the Osteopathic Field*. Handspring Publishing.

Lubbock, J. (1887). *The Pleasures of Life. Part II*. Leopold Classic Library 2016.

MacIntosh, J. E., & Bogduk, N. (1991). "The attachments of the lumbar erector spinae." *Spine, 16*(7), 783–792.

Maitland, J. (2001). *Spinal Manipulation Made Simple*. North Atlantic Books.

Martin, J. (1977). "In activity therapy, patients literally move toward mental health." *The Physician and Sportsmedicine, 5*(7), 84–89.

Mattern, S. P. (2013). *The Prince of Medicine: Galen in the Roman Empire* (1st ed.). OUP Oxford.

McGill, S. M. (2001). "Low back stability: From formal description to issues for performance and rehabilitation." *Exercise and Sport Sciences Reviews, 29*(1), 26–31.

McKeon, P. O., Hertel, J., Bramble, D., & Davis, I. (2015). "The foot core system: A new paradigm for understanding intrinsic foot muscle function." *British Journal of Sports Medicine*, *49*(5), 290.

McKeown, P. (2015). *The Oxygen Advantage: The Simple, Scientifically Proven Breathing Technique That Will Revolutionise Your Health and Fitness*. Piatkus.

Merriam-Webster online dictionary. (2022). https://www.merriam-webster.com/.

Merton, T. (1966). *Conjectures of a Guilty Bystander*. Bantam Doubleday Dell Publishing Group Inc 1994.

Mikkelsen, K., Stojanovska, L., Polenakovic, M., Bosevski, M., & Apostolopoulos, V. (2017). "Exercise and mental health." *Maturitas*, *106*, 48–56.

Miller, J. C. (2004). *The Transcendent Function: Jung's Model of Psychological Growth Through Dialogue With the Unconscious*. State University of New York Press.

Moseley, L., & Butler, D. (2013). *Explain Pain* (2nd ed.). Noigroup Publications.

Muscolino, J. E., & Cipriani, S. (2004). "Pilates and the 'powerhouse'—II". *Journal of Bodywork and Movement Therapies*, *8*(2), 122–130.

Myers, T. (2020). *Anatomy Trains: Myofascial Meridians for Manual Therapists and Movement Professionals* (4th ed.). Elsevier.

Naschitz, J. E., Mussafia-Priselac, R., Kovalev, Y., Zaigraykin, N., Slobodin, G., Elias, N., & Rosner, I. (2006). "Patterns of hypocapnia on tilt in patients with fibromyalgia, chronic fatigue syndrome, nonspecific dizziness, and neurally mediated syncope." *The American Journal of the Medical Sciences*, *331*(6), 295–303.

Nestor, J. (2021). *Breath: The New Science of a Lost Art*. Penguin Life.

NICE. (2019). "NICE guidelines. Urinary incontinence and pelvic organ prolapse in women: management." https://www.nice.org.uk/guidance/ng123.

Nietzsche, F. (1901). *The Will to Power (Penguin Classics)*. Penguin Classics; Translation edition 2017.

Nietzsche, F. (1908). *Ecce Homo*. Oxford World's Classics 2009.

Nixon, P. G. (1994). "Effort syndrome: Hyperventilation and reduction of anaerobic threshold." *Biofeedback and Self-Regulation*, *19*(2), 155–169.

Noland, K. (1988). "Context." Transcript of a speech at University of Hartford, March, 1988 at symposium *The Bennington Years*.

Nygaard, I. E., Thompson, F. L., Svengalis, S. L., & Albright, J. P. (1994). "Urinary incontinence in elite nulliparous athletes." *Obstetrics and Gynecology*, *84*(2), 183–187.

Ovid. (800). *Metamorphoses*. Penguin Classics 2004.

Phillips, S., Mercer, S., & Bogduk, N. (2008). "Anatomy and biomechanics of quadratus lumborum." *Proceedings of the Institution of Mechanical Engineers. Part H, Journal of Engineering in Medicine, 222*(2), 151–159.

Pickering, M., & Jones, J. F. X. (2002). "The diaphragm: Two physiological muscles in one." *Journal of Anatomy, 201*(4), 305–312.

Pico Della Mirandola, G. (1496). *Oration on the Dignity of Man*. Gateway Editions 1996.

Plaskin, G. (1984). *Horowitz: A Biography of Vladimir Horowitz*. Smithmark Pub; Reprint edition.

Polemnia, A. (2007). "Intentional control of motor-respiratory coordination." *Journal of Sport & Exercise Psychology, 29*, S51.

Poliquin, C. (1997). *Poliquin Principles*. Poliquin Performance Centers 2006.

Pope, M. H., Goh, K. L., & Magnusson, M. L. (2002). "Spine ergonomics." *Annual Review of Biomedical Engineering, 4*, 49–68.

Rahmani, N., Mohseni-Bandpei, M. A., Salavati, M., Vameghi, R., & Abdollahi, I. (2018). "Normal values of abdominal muscles thickness in healthy children using ultrasonography." *Musculoskeletal Science & Practice, 34*, 54–58.

Raja, S. N., Carr, D. B., Cohen, M., Finnerup, N. B., Flor, H., Gibson, S., Keefe, F. J., Mogil, J. S., Ringkamp, M., Sluka, K. A., Song, X.-J., Stevens, B., Sullivan, M. D., Tutelman, P. R., Ushida, T., & Vader, K. (2020). "The revised International Association for the Study of Pain definition of pain: Concepts, challenges, and compromises." *Pain, 161*(9), 1976–1982.

Remnick, D. (1999). *King of the World: Muhammad Ali and the Rise of an American Hero*. Picador.

Rolf, I. P. (1977). *Rolfing: Re-establishing the Natural Alignment and Structural Integration of the Human Body for Vitality and Well-Being*. Inner Traditions Bear and Company; 1st Healing Arts Press Ed edition 1992.

Salvatore, S., Serati, M., Laterza, R., Uccella, S., Torella, M., & Bolis, P.-F. (2009). "The impact of urinary stress incontinence in young and middle-age women practising recreational sports activity: an epidemiological study." *British Journal of Sports Medicine, 43*(14), 1115–1118.

Saranam, S. (2008). *God Without Religion: Questioning Centuries Of Accepted Truths*. Simon and Schuster.

Saunders, S. W., Rath, D., & Hodges, P. W. (2004). "Postural and respiratory activation of the trunk muscles changes with mode and speed of locomotion." *Gait & Posture, 20*(3), 280–290.

Schuenke, M., Schulte, E., Schmacher, U., & Johnson, N. (2020). *General Anatomy and Musculoskeletal System (THIEME Atlas of Anatomy)* (3rd ed.). Thieme Medical Publishers Inc.

Sir John Lubbock. (1887). *The Pleasures of Life. Part II*. Leopold Classic Library 2016.

Smith, M. D., Russell, A., & Hodges, P. W. (2006). "Disorders of breathing and continence have a stronger association with back pain than obesity and physical activity." *The Australian Journal of Physiotherapy, 52*(1), 11–16.

Stecco, C. (2014). *Functional Atlas of the Human Fascial System* (W. I. Hammer, Ed.). Churchill Livingstone.

Steele, V. (2003). *The Corset: A Cultural History*. Yale University Press.

Steinacker, J. M., Both, M., & Whipp, B. J. (1993). "Pulmonary mechanics and entrainment of respiration and stroke rate during rowing." *International Journal of Sports Medicine, 14 Suppl 1*, S15–9.

Sueki, D. G., Cleland, J. A., & Wainner, R. S. (2013). "A regional interdependence model of musculoskeletal dysfunction: Research, mechanisms, and clinical implications." *The Journal of Manual & Manipulative Therapy, 21*(2), 90–102.

Terry, P. C., Karageorghis, C. I., Curran, M. L., Martin, O. V., & Parsons-Smith, R. L. (2020). "Effects of music in exercise and sport: A meta-analytic review." *Psychological Bulletin, 146*(2), 91–117.

Thompson, D. W. (1917). *On Growth and Form*. Cambridge University Press; Abridged edition 2014.

Tiberio, D. (2019). *Personal communication*.

Tsao, H., & Hodges, P. W. (2007). "Immediate changes in feedforward postural adjustments following voluntary motor training." *Experimental Brain Research, 181*(4), 537–546.

Tzu, L. (400 BCE). *Tao Te Ching (Penguin Classics)* (D. C. (Translator) Lau, Ed.). Penguin Classics; Reprint edition 2003.

Veres, S. P., Robertson, P. A., & Broom, N. D. (2010). "The influence of torsion on disc herniation when combined with flexion." *European Spine Journal: Official Publication of the European Spine Society, the European Spinal Deformity Society, and the European Section of the Cervical Spine Research Society, 19*(9), 1468–1478.

Verne, J. (1864). *Journey to the Centre of the Earth (Oxford World's Classics)*. OUP Oxford; Reprint edition. 2008.

Vleeming, A., Mooney, V., & Stoeckart, R. (2007). *Movement, Stability & Lumbopelvic Pain: Integration of Research and Therapy* (A. Vleeming, V. Mooney, & R. Stoeckart, Eds.; 2nd ed.). Churchill Livingstone.

Ward, S. R., Kim, C. W., Eng, C. M., Gottschalk, L. J., Tomiya, A., Garfin, S. R., & Lieber, R. L. (2009). "Architectural analysis and intraoperative measurements demonstrate the unique design of the multifidus muscle for lumbar spine stability." *The Journal of Bone and Joint Surgery. American Volume, 91*(1), 176–185.

Wells, H. G. (1945). *Passionate Friends: Adapt or Perish, Now As Ever, Is Nature's Inexorable Imperative*. Wentworth Press 2016.

Wesnes, S. L., Rortveit, G., Bø, K., & Hunskaar, S. (2007). "Urinary incontinence during pregnancy." *Obstetrics and Gynecology, 109*(4), 922–928.

Willardson, J. M. (2007). "Core stability training: Applications to sports conditioning programs." *Journal of Strength and Conditioning Research, 21*(3), 979–985.

Wirth, K., Hartmann, H., Mickel, C., Szilvas, E., Keiner, M., & Sander, A. (2017). "Core stability in athletes: A critical analysis of current guidelines." *Sports Medicine (Auckland, N.Z.), 47*(3), 401–414.

Woodham, M., Woodham, A., Skeate, J. G., & Freeman, M. (2014). "Long-term lumbar multifidus muscle atrophy changes documented with magnetic resonance imaging: A case series." *Journal of Radiology Case Reports, 8*(5), 27–34.

Zemeckis, R. (1994). *Forest Gump*. Paramount Pictures.

Index

Acknowledgments

It is a rare moment when one is asked to write a book. For me this moment combined gratitude with terror. And so it was that I began my journey to the core.

What lies in your hands is the result of my journey of discovery. My hope is that you will take as much pleasure in reading this book as I had in its writing. My initial trepidation at this sometimes-controversial topic has been replaced with a fondness that I hope is conveyed by the words chosen.

That I dare to offer this book to you stems not from some misguided idea that I now know all there is to know. This book and the information within it are offered with a feeling of satisfaction gained from the pursuit of knowledge. A pursuit that is far from over, and one I hope continues for writer and reader alike.

In the writing of this book there arose many difficulties, perplexities, and puzzles. It was at those moments of dark confusion that a glimmer of understanding became possible. It is my hope that the reader, in the same way, also finds moments to challenge and so to learn.

It is now down to you, the reader, to use the ideas presented here as a catalyst to your own discoveries. I urge you not to be confined by my own opinions and limits but to continue to explore. This is not a book to follow as some kind of dogmatic truth to be adhered to. Instead, hold the ideas lightly and allow the spark of imaginative movement to unfold into ceaseless exploration.

While "author" is a singular term, writing a book is a collaborative process of gathered talents, each of whom I sincerely thank.

I am particularly grateful to the catalyst, Jon, of Lotus Publishing, for the initial idea that began this journey, and for his willingness to publish the resulting work.

Thanks too go to Scott and Madelaine, the movement explorers, for their willingness to be photographed. Thanks too to Joel, the skilled photographer able to capture movement in the still image.

I am always in debt to the invaluable support of James Earls, whose kind words can be found within these pages. It is with intense pride that I collaborate with James at Born to Move to inspire students around the world.

A special mention goes to Malou and Ezra, who encapsulate the central importance of fun and joy in all movement.

And then there is Jen, without whom I would not be.

Lastly, thanks to all my clients who, often from no fault of their own, missed or cancelled appointments, thus allowing me stolen moments to read and scribble.